Macmillan Building and Surveying Series
Series Editor: **Ivor H. Seeley**
Emeritus Professor, The Nottingham Trent University

List continued overleaf

List continued from previous page

Macmillan Building and Surveying Series
Series Standing Order
ISBN 0–333–71692–2 hardcover
ISBN 0–333–69333–7 paperback
(outside North America only)

You can receive future titles in this series as they are published by placing a
standing order. Please contact your bookseller or, in the case of difficulty, write
to us at the address below with your name and address, the title of the series
and the ISBN quoted above.

Customer Services Department, Macmillan Distribution Ltd
Houndmills, Basingstoke, Hampshire RG21 6XS, England

BUILDING SURVEYS, REPORTS AND DILAPIDATIONS

IVOR H. SEELEY

BSc (Est Man), MA, PhD, FRICS, CEng, FICE, FCIOB, MIH

MACMILLAN

First published 1985 by
THE MACMILLAN PRESS LTD
Houndmills, Basingstoke, Hampshire RG21 2XS
and London
Companies and representatives
throughout the world

ISBN 0–333–40209–X hardcover
ISBN 0–333–36859–2 paperback

A catalogue record for this book is available
from the British Library.

13 12 11 10 9   8   7   6
03 02 01 00 99 98 97

Printed in Malaysia

'The Moving Finger writes; and, having writ,
Moves on; nor all thy Piety nor Wit
Shall lure it back to cancel half a line,
Nor all thy Tears wash out a word of it.'

Rubaiyat of Omar Khayyam
(Persian mathematician born in about the year 1050)

Contents

Preface

This book is aimed at professionals involved in the preparation of structural and other types of building report, schedules of condition, schedules of dilapidations and proofs of evidence. It is believed that it will also be of considerable value to student surveyors on degree courses in building surveying and estate management and to those studying for the relevant professional examinations.

The method of undertaking building and structural surveys, including the equipment required, the sequence of survey operations and the recording of the relevant data are described and illustrated. The methods of preparation and presentation of reports for a variety of purposes are examined, with an appraisal of the divergent views being expressed and the surveyor's professional liability.

Then follows a study of the principal building defects, illustrating their nature, method of identification at the survey stage, probable consequences and means of rectification.

Examples are included of a structural survey of a domestic building and a structural report of an industrial and commercial property, both accompanied by extensive explanatory notes to enhance their value to the reader. Alternative approaches to structural report preparation are also illustrated.

A comparative study of structural and technical reports, schedules of condition and proofs of evidence highlights their main features and they are each illustrated by case studies. The role of the surveyor as an expert witness and arbitrator is detailed as these are becoming increasingly important functions.

Finally, a study is made of the nature and scope of dilapidations and their legal implications, together with a number of related aspects with which the surveyor may be concerned. These include covenants to repair and the remedies for their breach, standards of repair, and such cognate matters as party walls, dangerous buildings and fixtures. This proceeds logically into the method of preparation of schedules of dilapidations, supported by several practical examples, including the use of a Scott schedule in settling dilapidations claims.

Nottingham
Spring 1985

IVOR H. SEELEY

Acknowledgements

A debt of gratitude is owed to the many professionals with whom I have come into contact over the years, and I am much the wiser for their opinions and views.

I have drawn freely on a number of publications of the Royal Institution of Chartered Surveyors on structural surveys and reports, which offer much sound and authoritative advice. West's comprehensive book on the law of dilapidations provides an invaluable source of reference and was of considerable assistance to me in writing chapter 7.

Grateful thanks are due to Macmillan Publishers Ltd for kind permission to quote from *Building Maintenance* and to the Ellis School of Surveying and Building, Worcester, for consent to use some of the concepts contained in course material prepared by the author for the School in years past. The Architectural Press Ltd kindly agreed to the inclusion of appendix A and appendix B.

Much helpful and friendly advice was received from John Winckler of the publishers, for which I am most grateful. Mrs E. D. Robinson prepared the typescript in her customary efficient and expeditious way, and my wife continued to give me full encouragement and support in my book-writing activities.

List of Tables

List of Figures

Table of Statutes

Table of Cases

1 Building Surveys

This chapter provides an introduction to the form and purpose of building surveys and attempts to clarify some of the confusion that has arisen in connection with them. The contractual arrangements between the surveyor and the client are examined together with the duties and obligations of surveyors when carrying out surveys of buildings.

Form and Purpose of Building Surveys

Building Society Surveys

Building surveys can be undertaken in varying degrees of detail and to serve a number of different purposes. Most house purchasers first become aware of such surveys when they approach a building society to obtain a mortgage and the society commissions a surveyor to establish the condition and value of the property and to decide the sum which the society is prepared to loan to the prospective purchaser. Each society tends to prefer its own survey forms, resulting in wide variations in format and content, and more recently some societies have introduced the RICS house buyers' report and valuation form with various modifications.

Until quite recent times the reports emanating from these surveys were regarded as confidential to the building society and were not communicated to the prospective purchaser, even although he paid the survey fee. The situation changed dramatically when the Abbey National Building Society decided to make copies of its reports available to intending mortgagees. This gave rise to doubts in the minds of some surveyors as to whether their principal duty was to the society or the applicant, although the author believes that it must rest with the society.

House Buyers' Report and Valuation

In 1981, the Royal Institution of Chartered Surveyors produced a House Buyers' Report and Valuation Inspection Report Form[1] for use by surveyors in determining the general state of repair of the property being surveyed and giving valuation advice. The form was revised slightly in 1984. It is not a structural survey as generally understood but is reasonably comprehensive and incorporates information under the following heads, and includes many caveats.

General
  (1) Name and address of client.
  (2) Address of property inspected.
  (3) Date of inspection.
  (4) Weather.
  (5) Tenure and a note of tenancies (where applicable).
  (6) Rateable value (from oral enquiry).
  (7) Description of the property, encompassing its type and age, unusual factors regarding location, such as its remoteness, or if situated on a steep hill, and a brief description of the accommodation including garage(s) and outbuilding(s).

Externally
  (8) Chimney stacks, flashings and soakers, as observed from the ground.
  (9) Roofs, with attention being drawn to roof slopes or flat areas which are not visible and have therefore been excluded; also roof spaces where there are access hatches reasonably accessible.
 (10) Parapets and parapet and valley gutters.
 (11) Gutters, downpipes and gullies (unless raining at the time of the inspection, it might not be possible to state whether or not the rainwater fittings are watertight).
 (12) Main walls (examined from ground level and foundations are not opened up for examination).
 (13) Damp-proof courses and sub-floor ventilation.
 (14) External joinery, including window and door frames, examined as far as possible.
 (15) Exterior decorations and paintwork, with the general condition only being noted.

Internally
 (16) Ceilings, walls and partitions, inspected from floor level but without moving furniture and wall hangings.
 (17) Fireplaces, flues and chimney breasts, although the condition of flues or presence of flue liners is omitted.
 (18) Floors: the surface of all uncovered floors is inspected as far as practicable, but fixed floorboards are not lifted, and accessible corners of fixed coverings will be lifted sufficiently to identify the nature of the finish beneath.
 (19) Dampness: damp meter readings are made where appropriate at the external and internal walls and floors, but without moving heavy furniture, fixtures and fittings.
 (20) Internal joinery, including doors, staircases and built-in fitments – general comments only.
 (21) Internal decorations – general comments only.
 (22) Cellars and vaults – general comments only.
 (23) Woodworm, dry rot and other timber defects (defects revealed by the

examination of the structure, but excluding those areas of the building which are covered, unexposed or not readily accessible).

(24) Thermal insulation (overall comment is made in connection with visible areas, but it may not be possible to verify the information given or the condition of the material).

Services (inspected visually where accessible, tests are not applied and comments on the general condition only are given).

(25) Electricity.

(26) Gas (if connected).

(27) Water, plumbing and sanitary appliances.

(28) Hot water and central heating (internal heating appliances normally require a flue liner, but a visual inspection does not always reveal that one has been fitted).

(29) Underground drainage (a visual inspection is made where it is possible to locate and raise the covers of inspection chambers).

General

(30) Garage(s) and outbuilding(s): comments are restricted to important defects; other structures, such as swimming pools and tennis courts, are not inspected.

(31) The site: general reference is made and only significant defects in boundary fences, walls, retaining walls, paths and drives are reported, with reference to such features as flooding and tree roots included where applicable.

(32) Building regulations, town planning, highways, statutory, mining and environmental matters.

(33) Summary and recommendations (the main defects are indicated with a note as to whether these defects are normally found in property of this type and age).

(34) Valuation (advice is given on the current market value of the property at the date of inspection, taking into account its repair and condition, but excluding carpets, curtains and the like).

(35) Limitations: the report is based on the following assumptions.

    (i) that no high alumina concrete or calcium chloride additive or other deleterious material was used in the construction of the property;

    (ii) that the property is not subject to any unusual or especially onerous restrictions, encumbrances or outgoings and that good title can be shown;

    (iii) that the property and its value are unaffected by any matters which would be revealed by a local search and replies to the usual enquiries, or by any statutory notice, and that neither the property, nor its condition, nor its use, nor its intended use, is or will be unlawful; and

    (iv) that inspection of those parts which have not been inspected would neither reveal material defects nor cause the surveyor to alter the valuation materially.

It will be apparent that a survey carried out in accordance with the RICS inspection report form, and the various directives incorporated in it, provides an effective overall general picture of the property, but that many important details are omitted. In many circumstances a prospective purchaser would be better served by the receipt of a more detailed survey of the type described under 'Structural Surveys'. Many building societies have reproduced the RICS form, omitting only the limiting notes and the marginal comments, and they sometimes use it for types of property for which the RICS admits that the approach is unsuitable.

Flat Buyers' Report

In late 1983, the RICS produced a Flat Buyers' Report Form[2] as the house buyers' report form published in 1981 deliberately excluded flats because of the special problems encountered in surveying them and, in particular, the difficulty in reporting on only part of a building, problems with access, and the existence of common parts and services.[3]

Roof inspection is covered in considerable detail as this is particularly important in the case of flats. If the only access is through another flat, the purchaser is responsible for ensuring that this access is available or he must accept the restricted nature of the report. In the latter event, the surveyor should emphasise the possible consequences, as the roof is one of the most costly elements in which to rectify defects. If the surveyor has to make a second visit, this will generally involve an additional charge. In the case of flats in large blocks, the scheme provides for a sample survey only, particularly with regard to the roof space. Common parts and stairways are limited to the main or nearest approach.

The drainage system, being common to a number of flats, is excluded from the inspection. Services are difficult to survey and are subdivided into those contained within the flat, and central systems encountered in large blocks where general comment and warnings are considered sufficient. Otherwise they are excluded.

Dampness is another difficult aspect and particular attention is required at the lowest or basement flat. Often when an upper flat is being inspected, the surveyor will not be able to gain access to any other parts of the building, as they are contained within other ownerships. With regard to older converted buildings, a damp-proof guarantee will be required, and strong warnings given if this is not available. It is important that the solicitor checks on this information and reports on the validity of the relevant documents. Guidance is also required on the impact of fire regulations and water board regulations.

Management problems may also occur. Leases are rarely available for checking by the surveyor. When the surveyor has presented his report, the solicitor will check responsibility for effective repair and management. Some of the deficiencies found are difficult to rectify, particularly in the case of converted flats, such as the lack of timber guarantees, damp-proof guarantees and planning permission, and difficulties over building regulations and fire regulations.[3]

Structural Surveys

General characteristics With structural surveys it is essential that the client's requirements should be clearly established at the outset, as these can vary widely with the type of building and its condition and between clients. The term 'structural survey' has been defined by the RICS[4] as "the inspection of visible, exposed and accessible parts of the fabric of the building under consideration."

In the case of commercial and industrial buildings, it is particularly important to carefully identify and evaluate the client's requirements as it is often impractical and unnecessary to inspect and report in as much detail as that needed for residential buildings. The client frequently requires a broad appraisal of the condition of the building and general advice as to the suitability of the building for the intended purpose. Nevertheless, clients' expectations will vary widely and must be clarified at the briefing stage. Where the client requires advice on the suitability of a building for a specific use, it is sometimes better to undertake a feasibility study as a separate brief.[4]

A RICS Practice Note on structural surveys of residential property[5] is aimed primarily at the inspection of property for intending purchasers or for surveyors advising clients on the structural condition and state of repair of residential properties. The practice note lists the surveyor's responsibilities in this type of assignment as follows:

(1) assessing the client's needs;
(2) determining the extent of the investigations to be made and obtaining instructions from the client for any additional services required;
(3) undertaking the survey of the property in the form required by the client to establish the condition of the property and reporting to him in the detail and format necessary to provide him with a balanced professional opinion; and
(4) complying at all times with the agreed instructions which form the contract between the client and the surveyor.

While a house buyer's report is relatively quick and inexpensive and includes a valuation, a structural survey is expensive and involves a great deal of time in looking for minor defects. A structural survey should however concentrate firstly on the structure and then deal with the less significant items later, so that major deficiencies are adequately highlighted.[6] It has been suggested that a better and more sensible approach would be for the vendor to commission the survey and make the report available to all prospective purchasers to avoid duplication of effort and expense.

Practical issues Any structural survey is invariably incomplete, in that some parts of the building are not visible. The method of assembly of the component parts of the structure conceals much of their nature, quality and condition. Further concealment results from furnishings in an occupied building.[7]

The following examples illustrate ways in which some of these limitations may be overcome. Furnishings and floor coverings may be moved and floorboards lifted. Holes may be drilled in external walls to enable cavities to be inspected by a fibre-optic viewer. The ground may be excavated to expose drains or foundations. Samples of plaster and mortar may be removed for chemical or other analysis and examination. Tiles or slates may be stripped from roofs to expose the construction and materials below. Ladders may be brought to the site to enable concealed valleys and flat roofs to be investigated, and ladders and/or scaffolding provided for the examination of the upper faces of walls. Calculations can be made to assess the strength and adequacy of beams and/or joists, and pressure and flow tests carried out on underground drains.[7]

As a preliminary or an alternative to exhaustive tests, a surveyor may look for signs that give warning of hidden defects. He may then either recommend to the client that more detailed tests be undertaken or give the client notice of the risks that cannot be resolved in the absence of such tests. In almost every structural survey, the extent of tests and opening up is limited by physical factors and/or financial constraints. A prudent surveyor will notify the client of these limitations. An exhaustive structural survey is very time-consuming and, because the surveyor's fee is normally proportional to the time spent, it will also be expensive.[7]

Where the evidence obtained does not point conclusively to the need for repair, the surveyor must discuss at some length the need for repair work and the risks and consequences of postponing it. These aspects cannot be included in a normal valuation report. Wilde[7] has described how most building societies provide little space on their report forms for matters relating to structural condition, and so it is apparent that they are not requiring detailed information on them. Where defects are identified within the limitations of the survey, then it is clear that the building society requires the surveyor to take them into account in making his valuation and to refer to them only if repairs are required. A prospective borrower requiring more detailed structural information must commission his own survey.

Problems of implementation The comprehensive structural survey is generally regarded as the ultimate in the investigation of the structure of a building. Some eminent building surveyors consider that the term 'structural survey' should be confined to surveys undertaken by structural engineers to cover such aspects as foundations and the structures of framed buildings, and that the majority of detailed building surveys could more appropriately be described as 'reconnaissance surveys'. This latter terminology does not seem entirely appropriate as it suggests that the survey is very much of an exploratory nature.

It is worth noting that more than 90 per cent of claims investigated by the National House Building Council related to the failure of sub-structures, often where houses had been built prematurely on unsuitable ground with insufficient or no subsoil analyses. One of the worst examples, as described by Melville,[6] was the use of shale infilling in north-east England. The shale was extracted from tips before it had time to weather and before constituents such as gypsum and alum were leached out, resulting in extensive expansion. The National House Building Council subsequently banned the use of shale and devised remedial measures, including the use of suspended floors.

Melville[6] has formulated the following four useful guidelines relating to the carrying out of building surveys:

(1) Structural elements of buildings require careful scrutiny. For example, it is necessary to determine whether the roof is adequately supported and braced, that all essential structural members are in position, that the external walls do not bow or lean outwards and that the interior structure of the house is sound. If there is evidence of movement in the main walls and, in particular, below the line of the damp-proof course, this suggests foundation problems and necessitates thorough examination.

(2) All appropriate lines of enquiry concerning the property must be pursued, including pressing solicitors to obtain as much information as possible from vendors.

(3) Statements about repair or disrepair should be kept in correct perspective. For instance, if a valuer inserts in a building society valuation, under the heading of condition, that the top four courses of a chimney stack need repointing and that a gutter joint is defective, the reader of the report may assume that because the valuer has been so thorough in reporting these particular defects, that these are the only deficiencies in the property. He may further assume that there is no need to have a structural survey or make any provision for other repairs. This policy of giving detailed descriptions of relatively minor repair works should only be adopted if it is to be applied consistently to all parts of the building.

(4) If a surveyor finds a fault, he is under a duty to trace it to its source. A general practice surveyor can adopt the approach that he is under no obligation to pronounce on matters outside his area of expertise. He does however then have a duty to warn his client and he could discharge this duty by stating 'I do not understand this problem and further advice should be sought'. The valuer or general practice surveyor in this situation should advisably recommend that a full survey be carried out or that the applicant obtains specialist advice. Unfortunately, a large proportion of claims received by RICS Insurance Services arise from statements made by surveyors, largely to reassure clients, where insufficient information is provided.

Needs of the Client

The client faces a number of problems concerned with identifying the type of survey that best meets his needs, who should undertake it, and finally the ability to effectively interpret the report when it is received. The surveyor normally starts from the premise that a client wishes to purchase a property, regardless of its faults unless they are exceptionally serious ones. The surveyor's primary task is often to identify the terms and conditions under which it will be safe and reasonable for the client to purchase the property of his choice.[8]

A surveyor should present all the relevant facts and the client then decides whether or not to purchase the property on the basis of the information provided. The surveyor has an obligation to make sure that the client is aware of the risks he is taking.

The surveyor should keep his report as free from technical terms as possible but the inclusion of some of the more generally used terms is inevitable as it is difficult to avoid them. One solution is to provide a glossary of technical terms as an appendix to the report.

The surveyor will also need to report on external factors which could influence the client's decision to purchase the property. These factors can encompass a wide range of matters from obnoxious fumes from a factory in a nearby road, to the use of a residential road as a connecting link between two radials and its designation by the Automobile Association for this purpose, resulting in a considerable volume of heavy traffic using the road, and extensive car parking on the road frontage by cars from a nearby garage or users of sports facilities in the district. Proximity to a clearance or redevelopment area is another deleterious factor. Serious defects in adjoining premises which could adversely affect the property being surveyed should be included in the report.

Where the client is a keen do-it-yourself man, he may require additional information and advice about the property, and particularly with regard to potential alterations. A common desire is to convert the existing lounge and dining room into a through room, and advice may be needed on the constructional work entailed.

Mortgage valuation reports generally require the surveyor to state whether the property is affected by subsidence or landslip. This is a vital aspect as the surveyor would be very unwise to recommend a client to purchase a property on which subsidence insurance will not be granted. There are also certain defects in buildings which can be justified on structural grounds but which can have a significant effect on value. For example, the use of tie rods as a lateral support to a flank wall is often sound building practice but is likely to worry potential purchasers. Underpinning can have a similar effect.[8]

Professions Undertaking Structural Surveys

Arguments as to which profession is best suited to carry out structural surveys — building surveyors, general practice surveyors or structural engineers — occupied a considerable amount of space in the correspondence columns of *Chartered Surveyor Weekly* in 1983 and 1984.

Following the receipt of building society valuation reports and house buyers' reports, some building societies are requesting supplementary reports from structural engineers and other specialists on what may be relatively minor faults in the property. Some believe that this stems from the fact that some surveyors appear to have inadequate knowledge and understanding of building construction and building defects. Others have gone further and point to a diminishing building construction content in some estate management degree courses as one of the root causes.

In general, chartered building surveyors can offer a much more comprehensive service than structural engineers. On the other hand, a building society is normally restricted to implementing its valuer's recommendations. Furthermore, the general public and some building societies have difficulty in distinguishing between a general practice surveyor and a building surveyor. However, it must also be accepted that a chartered surveyor in the general practice division, who is regularly involved in carrying out structural surveys and valuations for a wide range of clients over a broad spectrum of properties, cannot realistically be regarded as unsuited to undertake this class of work. He will have acquired a sound knowledge of buildings with the ability to diagnose faults and the appropriate remedial measures.

The case has been reported of one building society senior officer who stated that the requirement in the society's valuation report for a further report from a structural engineer arose from the terms of the society's insurance policy. However when the particular insurance company was contacted, it confirmed that a report of a chartered surveyor suitably experienced in the execution of structural surveys was acceptable without reservation.

There must, however, be occasions when the services of a structural engineer and/or a mechanical and electrical engineer are needed, particularly with large buildings and complexes incorporating major structural and services problems. One chartered building surveyor writing in *Chartered Surveyor Weekly* described how he recognised his own inadequacies on structural matters and subsequently qualified as a structural engineer. There is a great danger that the public will accept the surveyor as a complete expert unless his limitations are carefully explained to them.

Independent expert advice on services may be a necessity. For example, plant rooms are frequently deficient in conditioning-plant capacity, maintenance manuals and detailed electrical layout drawings. In some cases installations may even be electrically unsafe. Likewise, if a chartered surveyor considers it necessary to recommend that other advice on the structure should be sought, particularly in respect of unequal structural movement, then the advice of a chartered surveyor competent in this class of work or a structural engineer should be obtained, and this action will not undermine the professional standing of the inspecting surveyor.

It behoves all professionals to recognise their own limitations and to know when to call in specialists. Undoubtedly there have been occasions where reports by specialists have proved to be of no greater value than the more generalised comments in the original survey, as they can be equally qualified. With such a diverse profession as surveying, there is bound to be some overlap between the divisions, and it would be most regrettable if the divisional structure of the RICS should lead to internal professional jealousies and attempts to prevent qualified and experienced members offering their skills to the public.

Problems can arise where a chartered surveyor is commissioned to carry out a structural survey of a property and, in particular, to comment on cracks in the property. Where the cracks are old and cannot affect the structural condition of the property, it does the surveyor little credit if he then recommends that the services of a structural engineer should be obtained to comment on the structural condition.

Every surveyor or valuer inspecting property for mortgage purposes should be capable of recognising settlement, and be able to distinguish settlement from cavity wall tie corrosion, sulphate attack, thermal movement and shrinkage. A surveyor should also either be familiar with the soil conditions of the area in which he practises or obtain geological plans of the district. He should also be able to identify those instances where building movement should be referred to a specialist for a second opinion.

The Yianni case in 1981, which will be considered in detail later in this chapter, resulted in many valuation surveyors taking a more cautious approach towards property inspection, not only to protect their client's interest but also to protect themselves against possible accusations of professional negligence.

Contract with the Client

Client's Requirements

Before undertaking a building survey, it is imperative that the purpose, nature and scope of the survey be determined. As described earlier, a surveyor is often requested to prepare a building or structural survey of a dwelling on behalf of a prospective purchaser, and this may sometimes include a valuation. Another approach is the requirement to complete a *pro-forma* report form issued by a bank, building society, insurance company or other financial institution, and this inevitably includes a valuation for mortgage purposes and probably an estimate of the cost of rebuilding the property for insurance purposes. Reference has also been made earlier to the RICS House Buyers' Report and Valuation Inspection Report Form which includes provision for the insertion of a valuation. Hence it will be seen that in many instances a valuation of the property is part of the service required by the client, and a surveyor must be familiar with local property values in order to give this advice. Where specialised properties, such as hotels, theatres, supermarkets and marinas, are involved, it may be advisable to consult a practice specialising in these classes of property. In these circumstances it would be advisable to consult with the client, particularly if this could result in a higher fee.

The RICS Practice Note on Residential Surveys[5] lays down useful guidelines for surveyors to follow when taking instructions from a client. In many cases a member of the public contacting a surveyor about a survey of residential property will have insufficient knowledge to give clear instructions, and hence will normally require advice as to the type and scope of survey that is most suitable. Time spent clarifying points of doubt at this stage will avoid greater problems later.

Confirmation of Client's Instructions

Immediately agreement is reached, the surveyor should follow this up at once by confirming in writing his instructions and intentions to the client, covering all appropriate aspects such as the following:

(1) Nature of the instructions — whether structural survey, valuation, redevelopment appraisal or other assignment.
(2) Agreed date for survey or method of arranging an appointment.
(3) Statement of surveyor's intentions covering a whole range of relevant matters, which could incorporate:
    (i) extent of inspection of the main structure;
    (ii) extent of inspection of outbuildings and grounds;
    (iii) extent of opening up of structure;
    (iv) limitations on inspection where surfaces are covered by fitted carpets, wall fittings and the like;
    (v) extent to which heavy or fitted furniture will have to be moved;
    (vi) any limitation of liability in the form of a general exclusion clause often dictated by the surveyor's professional indemnity insurance policy;
    (vii) extent of enquiries to be made of local and statutory authorities;
    (viii) extent to which the surveyor will test drains, electrical, heating and other services;
    (ix) basis of fee to be paid and any expenses; and
    (x) whether a valuation or estimate of building costs is to be included and the nature of any limitations or reservations which will be applied.

The RICS Practice Note[5] describes how many firms of surveyors send a short covering letter to the client confirming the nature of the instructions, the date set for the survey and the fee. A book or pamphlet describing in detail both the extent and limitations of the surveyor's intentions is frequently enclosed with the letter. When instructions are received through a third party, the surveyor should ensure, as far as practicable, that his confirmation of instructions is forwarded in its entirety to the client.

A RICS Guidance Note[4] highlights the requirements of the *Unfair Contract Terms Act 1977*, whereby surveyors cannot generally exclude or limit their liability, except in so far as the term or notice is reasonable. This provision reinforces the need for a formal agreement and the simplest procedure is to request the client to sign a copy of the instructions. Surveyors' reports almost invariably contain some reservations relating to limitations of the inspection and items not tested. The RICS Guidance Note rightly discourages the indiscriminate use of caveats/reservations.

Contractual Relationship and Requirements

A contractual relationship is forged between the surveyor and his client, whereby he renders a service, normally the carrying out of an inspection and the preparation of a report in return for an appropriate payment from the client. A relationship also exists in tort between the surveyor and his client and this may also extend to any third parties who read and act on the report. If the surveyor wishes to avoid third party liability under the law of tort, he needs to include a specific disclaimer in the report.

Staveley and Glover[9] give examples of where a surveyor's report is used inappropriately for purposes other than those for which it was intended. For instance a client may use a current market valuation for fire insurance purposes or a specialised report as an aid to the sale of the property at a later date. Some clients also have special needs such as disabled persons. A surveyor should also take steps to advise elderly clients about features which could cause them problems and where a client has young children he should be warned of any hazards which could result from glazing to low windows, beside stairways and that provided for the full height of doors.

On occasions a client may wish to purchase a property with a view to altering and/or extending it, or changing its use. In these circumstances the surveyor should advise the client that he may need to obtain approval under Building and/or Town and Country Planning Regulations and that such consents are not given automatically. It is also possible that the property being surveyed has been altered or its use changed in the past, possibly without obtaining any necessary approvals and sometimes as part of a do-it-yourself activity. These cases need careful scrutiny by the surveyor to satisfy himself that no irregularity has occurred, and he must notify the client of his findings and their possible implications.

Fees

The basis for calculation of the surveyor's fee should be agreed with the client before the survey is commenced. Varying approaches are adopted in practice, ranging from a rate per room of accommodation to a proportion of the purchase price of residential properties. Both of these arrangements can operate unfairly against the client and a better approach is generally to assess the fee on the basis of an agreed hourly rate of time spent on the assignment, plus the cost of specified expenses. Where both a structural survey and a valuation are undertaken, the fee will normally equal the aggregate of the two separate fees that would have been chargeable had they been carried out as separate and distinct operations. Regard must also be paid to the need for the addition of travelling and other incidental expenses and the inclusion or otherwise of Value Added Tax.

There should be a clear and well-defined office procedure for recovering outstanding fees as it is important that they should not be permitted to remain unpaid indefinitely. In the final event and in exceptional cases, the surveyor can sue the client for non-payment of fees through the County Court under the small claims procedure.

Duties and Obligations of Surveyors

Negligence

The term 'negligence' has a wide interpretation in law and can be applied to many acts of omission, commission or carelessness, as described by Biscoe[10] as long ago as 1953. The general principle of the law of negligence is the failure to exercise that care which the particular circumstances demand. What amounts to negligence thus depends on the facts of each particular case, and the term does not therefore lend itself to a precise definition of universal application. It may consist of doing something which ought either to be done in a different manner or not done at all, or in omitting to do something which ought to be done.[10] It is the duty of the court to define the standard of care required, and this is founded on a consideration of the degree of care which would be observed by a prudent and reasonable man.

Professional Negligence

There is a duty to exercise special care in the case of a person engaged in a profession. In Halsbury's *Laws of England* it is stated that "The practice of a profession, from its nature, demands some special skill, ability, or experience and carries with it a duty to exercise to a reasonable extent, the amount of skill, ability and experience which it demands. If a person so practising fails to possess that amount of skill and experience which is usual in his profession, or if he neglects to use the skill and experience which he possesses or the necessary degree of care demanded or professed, he will usually be liable for breach of contractual duty."

The extent of the duty is honestly and diligently to use that care and skill which would be used at the time by other competent persons in the same profession. A surveyor will not, however, generally be held liable for injury resulting from a mere error of judgement on a difficult point, provided he can show that he exercised proper diligence in ascertaining all relevant facts before giving his opinion. If a surveyor acts in accordance with the practice accepted as proper by a responsible body of surveyors skilled in that particular art, he is not negligent merely because there is a body of opinion that takes a contrary view. A person should not undertake skilled work unless he is fitted to do so, and it is his duty to know whether or not he is so fitted.[11]

Negligence must be proved and the onus of proof lies on the party who alleges damage. Where, however, the negligence is of such an obvious and patent a nature that it speaks for itself, then the surveyor must assume the burden of satisfying the court that he applied with proper diligence that care, skill and experience which could have been expected from other qualified persons in similar circumstances.[10]

Extension of Liability

Cane[12] has described how the degree of responsibility of surveyors to their clients and third parties has increased significantly in recent years, both in monetary terms in excess of the rate of inflation and in the incidence of liability. The extension of liability appears to arise from three main sources:

(1) Increased market demand generated by a general increase in people's expectations. This has been coupled with surveyors undertaking more public relations promotion activities and thereby extending clients' expectations. Clients are coming to expect more from the professional at relatively lower cost in an increasingly commercial situation with government action resulting in competitive fees.
(2) Increased familiarity with legal remedies, whereby the public has been encouraged to take legal action to right their wrongs. The growth of the consumer protection movement has also accelerated the process.
(3) Insurance agreements have increased substantially the amount of money available for compensation. In addition poor clients can seek legal remedies with assistance through the legal aid scheme. Claims insurance is far more costly to administer than loss insurance. Nevertheless, as long as professionals consider that the price of the insurance premium, although high, is less than the value of the risk entailed, they will continue to pay premiums and to generate the funds to meet the claims,[12] and this was extended by the RICS compulsory professional indemnity insurance requirements in 1986.

Extent of Liability

A surveyor is liable to his client if he fails to perform what he has promised to do. He may also be liable, irrespective of what action he has promised, to those he injures through lack of using reasonable care. Before a surveyor can be required to pay compensation under this latter category, it is necessary to satisfy three criteria:

(1) that a duty of care existed;
(2) that he failed to use reasonable care; and
(3) that damage resulted.

The Limitation Act 1939 appeared to indicate that a party was barred from proceeding with an action after a prescribed period following the occurrence of the wrongful act or omission (six years in the case of property loss or damage and twelve years for breach of a contract under seal). It did not deal expressly with the case of a person who could not know that an injury or damage had occurred until after the expiration of the prescribed period. To remedy this injustice the courts developed the notion of fraud in the concealment of the injury or damage and held that, where it had been concealed wrongfully, time ran from when the plaintiff became aware of the wrongful act or omission.

Hence in the case of property damage, as for example with poorly constructed buildings, no right of action arises until the owner knows or could reasonably be expected to have known of the defects (*Anns v. Merton London Borough Council* (1977) 2 WLR 1042).

The giving of professional advice by a surveyor can have far-reaching consequences, where it is evident that the advice is likely to be acted on. He owes a legal duty of care, and if he is negligent in giving that advice he will be liable. This is illustrated in *Yianni v. Edward Evans & Sons* (1981) 3 WLR 843, where a building society commissioned a firm of surveyors and valuers to value a house for mortgage purposes. The prospective purchasers paid for the building society survey and were advised by the society to commission their own independent survey, but they declined to do so. The surveyors recommended that the maximum lending figure be applied and negligently failed to discover defects, of which the cost of repair exceeded the value of the house. The plaintiffs subsequently purchased the house on the basis of the surveyors' valuation. The surveyors must have realised that they would do this and were in consequence liable to them for the loss sustained.

Practical Examples of Professional Negligence

All relevant facts derived from a building survey must be incorporated in the report, together with the implications of these facts. A simple example will serve to illustrate this point. It is not sufficient merely to draw attention to the absence of air-bricks in a dwelling with a suspended timber ground floor. It must be made clear to the client that the omission of air-bricks can create conditions conducive to an outbreak of dry rot in the timbers because of the lack of ventilation, and the serious consequences that this can entail. It would be unreasonable to expect the client to infer this problem from a statement in the report confined to the absence of air-bricks. The surveyor should extend this item to include the condition of the floor timbers or, where inaccessible, to advise the client of the reasons for the failure to inspect, to express an opinion as to their possible condition and possibly conclude with a recommendation that the floor timbers should be inspected.[10]

Biscoe[10] described a report which stated that a property had a very shallow but satisfactory drainage system discharging into a sewer. During the description of the accommodation earlier in the report, mention was made of a water closet in the basement but no reference was made to the fact that this could not possibly be connected to the drainage system. This showed a serious deficiency in stating the determined facts and in drawing the necessary conclusions.

A very interesting and unusual law case relating to a building survey is that of *Last v. Post* in the High Court (1952). A surveyor was engaged to survey a house and during his inspection he examined the roof from inside the building and also with the aid of binoculars from the outside. He reported that the roof was of sound construction and the client purchased the property. It was subsequently discovered that the roof tiles were disintegrating at the heads as a result of efflorescence. An action was commenced and damages claimed to cover the cost of retiling the roof. Negligence was however denied on the grounds that a careful inspection of the roof had been made and that there was nothing to suggest that extra precautions should have been taken.

Lloyd-Jacobs J. in his judgement stated "that the defendant, who was of considerable standing as a surveyor, had no doubt followed his normal practice when inspecting the premises. Efflorescence was a rare phenomenon not present to the defendant's mind when he examined the roof and the defendant had genuinely and honestly thought that he was paying adequate attention to everything possible. The real explanation was that the defendant had failed to observe the deposit from the roof which must have been present on the flooring in some quantity at the time of his inspection. As efflorescence was so rare it would not be right to apply its detection as a standard of a competent surveyor, but he felt bound to hold that having failed to observe the deposit the defendant had failed in his duty. That did not mean that the defendant should have identified the phenomenon, but he should at least have expressed his report in such a way that the plaintiff would not have been led to believe that the roof was satisfactory and would not cause trouble and expense in the future."

It appears that in the absence of any deposit on the floor below the roof, the action would have failed. Had the deposit been observed, then the inference that the dust had come from the roof tiles would have been examined and the defective tiles reported on, or at least some cautionary words would have been used in order to make the client aware that there was a defect in the roof coverings.

Defects in Reports

It might be useful at this stage to examine some of the more common deficiencies which may occur in reports, in order to emphasise the diligence and care with which the surveyor must approach this assignment to avoid a possible action for negligence.

It is advisable to state clearly and concisely, at the beginning of the report, the agreed instructions and nature and scope of the report. A typical example of the introduction to the survey of a dwelling house could be as follows:

Ivy House, Martlesham

In accordance with your instructions, we have inspected the above property for the following purposes:

(1) Carrying out a structural and sanitary survey.
(2) Testing the drainage system.
(3) Testing the electrical, plumbing and heating installations.
(4) Determining whether the property is free from dry rot, wet rot and wood insect infestation.

Advice to the client must be sufficiently comprehensive for him to understand the nature of any defects and their likely implications. For example, a statement that 'the rear wall to the property is damp' adds nothing to what the client can see for himself. The statement gives no advice as to the cause or seriousness of the dampness or the cost or remedying it. The following represents a much more informative and professional approach.

'Dampness is penetrating the rear wall of the property because of soil being deposited above the horizontal slate damp-proof course to a height of about 250 mm (10 inches). In addition some of the bricks are flaking and porous.

'The dampness has caused wet rot to the skirting adjoining the rear wall and has also caused rotting of the ends of some of the floorboards in contact with the skirting. The wall plate and floor joists have been examined and found to be unaffected by either dry or wet rot.'

The report should then give a brief description of the work recommended to remedy the defects, followed by an approximate estimate of the cost of the remedial work. It would be prudent to warn the client of the likelihood of the defects becoming increasingly extensive, if they are not rectified immediately.

Other matters which are sometimes dealt with in insufficient detail are now briefly described, to assist the younger surveyor in avoiding some of the more obvious pitfalls.

(1) Treatment of the tops of chimney stacks when viewed from ground level may appear to indicate some limited repointing, whereas a closer examination might reveal that substantial rebuilding is required.

(2) A zinc flat roof covering nearing the end of its effective life and which has been covered with bitumen may already have permitted water penetration, with consequent damage to roof joists and boarding.

(3) Where an electrical installation has been renewed in the last few years, it should still be tested as it could contain defects.

(4) Where a new boiler has been installed in a hot water supply installation, a careful check should be made on the condition of the pipework as this is unlikely to have been replaced at the same time as the boiler.

(5) A close examination of flooring is imperative as the risk of rot or insect infestation is often considerable. The comments of an official referee concerning a report which read 'Flooring on the first floor appears to be in fair condition although we could only judge from superficial examination,' are worthy of note.

In his view, having found beetle in the woodwork and wet rot on the ground floor, any competent surveyor would have lifted one or two floorboards on the first floor. In this particular case it was practicable to take up floorboards and so to expose a second layer of floorboards underneath which were riddled with woodworm. Hence he was judged negligent in failing to do so. In many cases lifting floorboards is difficult where the property is occupied and the boarding is covered with fitted carpets or, worse still, parquet flooring which cannot be disturbed. The customary practice is, however, to take up floorboards for inspection wherever practical, and otherwise to advise that a further inspection be made when the house is vacant.

(6) In the case of pitched roofs with missing tiles or slates, a close inspection is needed to attempt to determine the condition of the battens and roof timbers below. If the surveyor is unable to ascertain their condition he must warn the client of the possible position and, in very bad cases, of the possibility of having to renew the roof or, at the very least, having to strip, rebatten and retile or reslate the roof.

(7) Dampness can be a major problem, particularly in low-lying areas or where there is a high groundwater level. In old properties without damp-proof courses and with the ground floors well below ground level, the surveyor's inspection must be thorough and his advice to the client very clear and precise. A surveyor who in this situation asserts that the dampness results from a leaking roof and broken windows, when in fact porous brickwork is absorbing the surface water in the absence of any effective waterproof membrane, is not giving adequate consideration to the prevailing conditions. This incorrect advice is compounded if he then advises the client to spend money on conversion works, without first ensuring that the building is made and will remain dry.

(8) Where a client is considering converting a property to provide living accommodation, first and foremost he needs to know whether the fabric of the

building is good enough to justify the expense of conversion. For example, it would not make economic sense to go to the expense of converting two smaller rooms into one and installing a bathroom and central heating unless the structure is sound and has a reasonable life span.[11]

Client's Instructions

As mentioned earlier in this chapter it is imperative that the client's instructions to the surveyor should be clear, precise and fully understood by both parties. It is not inconceivable that what is perfectly clear to the surveyor is not in the least understood by the lay client. It is the surveyor's first duty to determine exactly what is required and to explain this fully to the client.

The dangers that can arise from gross misunderstandings over the nature and scope of instructions are exemplified in *Sindcock v. Bangs* (1952) in which there was a claim for fees and a counterclaim for damages for negligence. The architect involved asked the defendant if a detailed survey or a general opinion on the property was required. The matter was urgent and the client stated that he required a general opinion. The property was purchased and it was subsequently discovered that there was settlement at one corner of the house and that dry rot and woodworm were present, and it was alleged that the architect had negligently and in breach of his duty failed to give warning of the defects in the property. Barry J. summing up, stated that the architect had been in the position of a watch dog for the client, who required professional advice to ensure that any offer he made for the house was reasonable and that any defects which might have a material affect on the property should have been reported as a matter of duty. At the initial survey there had been neither the time nor the opportunity to carry out a detailed examination of the structure and only a general opinion had been sought, but making all allowances for the circumstances he was satisfied that the opinion expressed by the architect was wrong and it amounted to negligence and breach of duty.[10]

This illustrates the dangers that exist in accepting vague and unspecific instructions. Whatever the scope of the instructions as they may finally be agreed, it is wise practice to confirm them in writing and in detail to the client and to obtain his agreement thereto before beginning the inspection. Biscoe[10] believed that the wider and less specific the instructions, then the more comprehensive must be the survey and the report. A typical request to look at a certain property and to say whether it is in order for a purchase to be made should not be dealt with lightly, and normally entails a detailed survey, including possibly obtaining specialist advice. It is wise to ask the client for clarification and to ask him whether he will authorise and obtain permission for you to uncover foundations, test drains, lift floorboards, obtain ladders and other equipment, and in addition to engage specialists to report on the various services installations.[10]

It is important when accepting instructions to ascertain the purpose for which the survey is required, particularly when a conversion is contemplated, as it could be that the client's proposals cannot be implemented on constructional and/or legal grounds. Hence the advisability of stating, in a preamble to the report, the purpose and scope of the instructions. It must always be borne in mind that however small the property to be surveyed and whatever the fee, the client is entitled to expect the same degree of care and skill as is required in the survey of an expensive office block, and if he suffers damage because that skill has not been exercised, then the courts will protect him.

A surveyor on inspecting a property has a duty to report honestly and genuinely on what he finds, even although he knows that his findings will not please a client who has already set his heart on the particular property. A cautiously framed report which avoids important issues lays the surveyor open to a charge of negligence. At the other extreme there is sometimes a temptation to report adversely on a property which could be 'full of trouble', and unless the condemnation is fully justified the client may subsequently find that he has suffered loss and could then take action against the surveyor.[10]

Measure of Damages

The case of *Philips v. Ward* (1956) All ER 874 dealt with the measure of damages payable when a surveyor was negligent in surveying a house. The plaintiff wished to purchase a house and farm and was attracted to a manor house in Sussex built in 1610. The surveyor inspected the property and reported that the house was of very substantial construction, with the best materials and workmanship lavished on it, but there were minor defects which he set out in the report. He valued the property at between £25 000 and £27 000 and in 1952 the plaintiff purchased the house and farm for £25 000. Unfortunately the surveyor had been negligent and did not notice that the timbers were badly affected by death-watch beetle and woodworm. The timbers were so rotten that it was necessary to replace the roof with new timbers and also timbers in wall plates and in the cellars, at a cost of £7000. The Official Referee held that the proper measure of damage was not the cost of repair but the difference between the value of the house and farm in the condition described in the report and the value as they should have been described, which he assessed at £4000.

The plaintiff appealed and Lord Denning ruled that the proper measure of damage is the amount of money which will put the plaintiff into as good a position as if the surveying contract had been properly fulfilled. He went on to say "If the defendant had carried out his contract, he would have reported the bad state of the timbers. On receiving the report, the plaintiff either would have refused to have anything to do with the house, in which case he would have suffered no damage, or he would have bought it for a sum which represented its fair value in its bad condition, in which case he would pay so much less on that account. The proper measure of damages is therefore the difference between the value in its assumed good condition and the value in the bad condition which should have been reported to the client."

A most extraordinary set of circumstances occurred in *London and South of England Building Society v. Stone* (1983) in the Court of Appeal. A couple wished to purchase a semi-detached house in Wiltshire in 1976, and approached the building society for a mortgage who, in turn, commissioned a valuer to give a general valuation and to report on whether the house would provide adequate security. The valuer reported that the house was worth £14 850 and would be good security for an advance of £12 800, and the society subsequently granted the applicants a mortgage of £11 800 and the couple entered into the usual covenants to pay interest and keep the property in good repair.

Shortly after purchase cracks appeared and doors ceased to fit. Consulting engineers in 1977 reported that the house, built in about 1961, was sited on an old and poorly filled quarry. The hillside was moving downward and the fill in the quarry was also sliding down, taking with it the foundations to the house. The house had either to be underpinned or demolished, with the shoring up of the other half of the semi-detached building. An insurance claim failed and the building society, as a gesture of goodwill, had the house repaired at an estimate of £14 000, although it finally cost £29 000.

Russell J. found the valuer negligent but was not persuaded that it was reasonable of the society to embark on the course that it did, admirable though its conduct was. He therefore made a deduction reducing the claim from £11 880 to £8800 to take account of the personal covenants of the mortgagees which the society had decided to waive. The society appealed and this was allowed in the Court of Appeal, illustrating the intricacies of the law in these cases.[13]

Avoiding Negligence in Structural Surveys

Ensom[14] describes how in an analysis of one thousand negligence claims relating to the surveying profession, 503 were in respect of structural surveys. It follows that if this category of claim could be reduced in number, it could lead to smaller premiums and reduce a significant area of complaint by the public. Some partners in private practice have stated that they will not undertake structural surveys because of the risks involved. Although this reaction is understandable, it cannot be in the best interests of either the profession or the public whom it serves.

This emphasises once again the importance of structural surveys being undertaken by persons with the necessary knowledge, skill and experience. RICS Insurance Services have stated that probably the greatest single cause of claims against surveyors arises from lack of agreement between the surveyor and his client, thus confirming the need for acceptance and confirmation of instructions as identified earlier in this chapter. Ensom[14] emphasises that a standard caveat inserted in a structural report offers no protection, if it was not known to the client when instructions were given. For example, where the surveyor is not going to test the services or comment on hidden timbers, then he should make this clear to the client before instructions are confirmed. It may be necessary in the report to make further reservations or to be more specific about those already made, but the main object of the instructions and their confirmation is to ensure that the client is not faced with an entirely changed situation from that which he expected.

Action on Receipt of a Claim

In the event of the receipt of a claim from a client, the surveyor is advised to start by examining his professional indemnity policy. It is likely to contain a clause stating that the surveyor on receiving a claim must not communicate with the claimant. In these circumstances, the correspondence and copies of all relevant documents must be sent to the insurers forthwith. Where, however, the policy does not prohibit the surveyor from communicating with the client, the policy will almost invariably require him to notify the insurance company that a claim has been received or is pending and that he is debarred from admitting liability or binding the insurers. It is, however, good practice to acknowledge the letter and to make an appointment to investigate the subject of the complaint.[14]

It may be that the client has not re-read the report after finding the defect. Had he done so he could have found it listed and described in the report, and in this case the matter could then be resolved fairly quickly. An allegation of negligence is a serious matter and justifies urgent attention by the senior partner, who should seriously consider arranging a meeting with the client. Where the allegation has some substance, the matter should be passed to the insurers, who will probably request an independent surveyor to inspect and to report.[14]

There is a danger that once a client finds a weakness in a report he is likely, either himself or through his adviser, to include every conceivable item where he feels in the least aggrieved, even although they would normally have found no place in the original claim. Where the insurers advise a settlement, this deserves serious consideration.[14]

Professional Indemnity Insurance

Nature of Professional Indemnity Insurance Policies

A professional indemnity policy is normally based on a 'claims made during policy period'. This means that insurers indemnify the practice in respect of claims made against it and notified to it during the policy period, and not for errors committed during that period. It therefore makes no difference if the negligence which is the subject of the complaint occurred several years previously or that it could be many years before the claim is settled — once notice has been given, the insurers will deal with it under the current policy year.

However, a policy formulated in this way can raise problems for the insured. When requesting insurance it is necessary to complete a proposal form which

requires full disclosure of all claims made against the practice and of all circumstances of which the partners are aware after reasonable enquiry and which
could result in claims being made against the practice. This is a fundamental point
as the proposal forms the basis of the contract of insurance and is contained
within it. Failure to make adequate disclosure will entitle the insurers to avoid
the policy and return the premium.

Wall[15] describes a professional indemnity insurance policy as a 'belt and
braces policy'. It protects the practice against claims, large and small, and safeguards years of goodwill. Without it the practice might founder. Furthermore, a
contract of insurance is one of the utmost good faith, with sound and confident
relationships between the insured and his insurers. Ideally one partner should
assume responsibility for the policy and any partner who is faced with a problem
which could involve the insurers should inform him immediately. In many
practices, professional indemnity insurance appears as a standing item on the
agendas of partners' meetings. It is vital that the insurers are aware of all the
activities carried out by the practice.

The policy is one of indemnity and thus normally indemnifies the practice in
respect of claims made against it arising from negligent acts, errors or omissions.
Hence the surveyor must have owed a duty of care, there must have been a
breach of that duty and the person making the allegation must have suffered a
loss.

The policy will not usually cover claims for breach of warranty or punitive or
exemplary damages. Unless an additional premium is paid, most policies do not
cover libel and slander or infringement of copyright, which could be significant
where a firm issues a brochure describing its activities or where a partner writes
articles or gives radio or television interviews. There is also unlikely to be cover
for claims made involving the dishonest misappropriation of client's monies
by the practice's employees or for claims involving loss of documents. Ideally,
the policy should be tailor-made to meet the precise requirements of the
practice and this is likely to involve lengthy discussions with insurance brokers
and insurers.[15]

Personal Injury

The surveyor who carries out a structural or valuation survey of a private house
may overlook or fail to report adequately on a defect in the property. If this
defect subsequently results in an accident, the surveyor is likely to face allega-
tions that any resultant personal injury or loss arose from his professional
negligence. It is prudent for surveyors engaged in almost any aspect of private
practice to give serious consideration to the possible consequences of a claim
for personal injury to third parties, and to the adequacy of their insurance
protection.[16]

The liabilities resulting from ordinary negligence can in most cases be covered
at relatively low cost under a public liability or third party liability policy. This

frequently takes the form of a section in a package policy covering most office insurance requirements. The limit of indemnity is likely to be £500 000 or £1 million for any one occurrence. The policy will normally exclude claims involving professional negligence. Liability for personal injury to third parties resulting from professional negligence must therefore be covered elsewhere. It is not unusual for a chartered surveyor to be covered for £250 000 under his professional indemnity policy and £1 million under his public liability policy.[16]

RICS Indemnity Insurance Scheme

Dann[17] describes how the financial survival of a practice may one day depend on whether a surveyor has adequate professional indemnity insurance. The RICS professional indemnity insurance scheme was introduced in 1976 to ensure that members could secure the protection they needed. Its principal objectives were:

(1) to produce the best available policy wording;
(2) to negotiate a favourable premium rating structure; and
(3) to enable the Institution to ensure, on behalf of members, that:
     (i)   trends in premium rating would be properly monitored and
     (ii) claims would be efficiently and sympathetically handled.

The professional indemnity policy wording negotiated for the scheme set a new standard for the surveying profession. It widened the cover in many important respects and severely restricted the circumstances in which the insurers could decline to deal with a claim for technical reasons. The RICS policy was further improved in 1984 by the introduction of new professional indemnity wording affording cover against all civil liability, subject to specified exceptions. This has the effect of including cover in circumstances where a liability is incurred in the course of professional practice without actually involving professional negligence.[17]

In each of the six years from 1977 to 1983, the cost of the claims made to RICS Insurance Services exceeded the premiums received by a considerable margin. 5163 claims were notified during this period of which 3428 had been resolved, in most cases to the entire satisfaction of the members involved. These results, it is claimed, have been achieved by a combination of able presentation by experienced staff in the claims department of RICS Insurance Services and fair, and often generous, treatment by the insurer's claims specialists. Experienced consultants are engaged where needed and there is a solicitor on the full-time staff of RICS Insurance Services.[17]

Since January 1986 it became compulsory for all chartered surveyors in private practice to have professional indemnity insurance for specified minimum limits. It must be borne in mind that surveyors in private practice have unlimited liability, and their financial ability to pursue their profession could depend on obtaining the correct professional indemnity insurance.

The need for professional indemnity insurance was highlighted in *Perry v. Phillips* (1982) 1 AER 1055 when Phillips, a surveyor, was held liable for negligence. In a structural survey he failed to discover a wall showing signs of bowing and miscellaneous other defects, and he was also held to be negligent for failing to identify a malfunction in the septic tank. The malfunction became evident when rainwater caused settlement problems. Other surveyors have admitted that they might have missed such a defect even in a full structural survey, partially since its appearance was at least particularly dependent on the weather.[18]

References

1. Royal Institution of Chartered Surveyors. *House Buyers' Report and Valuation Inspection Report.* 2nd edition (1984)
2. Royal Institution of Chartered Surveyors. *Flat Buyers' Report* (1983)
3. J. Hartley. RICS Flat buyers' report is completed. *Chartered Surveyor Weekly* (4 August 1983)
4. Royal Institution of Chartered Surveyors. *Guidance Note on Structural Surveys of Commercial and Industrial Property* (1983)
5. Royal Institution of Chartered Surveyors. *Structural Surveys of Residential Property: A Practice Note* (1981)
6. I. Melville. Surveys – perils, pitfalls and procedures. *Chartered Surveyor Weekly* (9 June 1983)
7. R. L. Wilde. Mortgage valuation or structural survey? *Chartered Surveyor Weekly* (23 June 1983)
8. I. Kemp. Structural surveys. *CALUS/ISVA Seminar* (1980)
9. H. S. Staveley and P. V. Glover. *Surveying Buildings.* Butterworths (1983)
10. G. Biscoe. Responsibility for structural reports. *Transactions of Royal Institution of Chartered Surveyors* (April 1953)
11. C. A. M. French. Professional negligence. *The Chartered Surveyor* (May 1960)
12. J. R. Cane. Professional negligence – responsibilities of practitioners. *Construction Surveyor* (April 1981)
13. Briefing Law Commentary. Assessing the cost of negligence. *Chartered Surveyor Weekly* (16 June 1983)
14. D. Ensom. Avoiding negligence in structural surveys. *Chartered Surveyor* (April 1980)
15. R. Wall. Professional indemnity. *Building Design* (10 June 1983)
16. RICS Insurance Services. Personal injury and professional negligence. *Chartered Surveyor Weekly* (28 July 1983)
17. RICS News. RICS indemnity insurance scheme. *Chartered Quantity Surveyor* (January 1984)
18. A. P. Lavers. *Towards Strict Liability.* Portsmouth Polytechnic, Department of Surveying (1983)

2 Structural Survey Procedures

General Objectives

As described in chapter 1, a prospective purchaser commissioning a building survey normally requires detailed advice on the structural condition of the property and an estimate of the cost of essential repairs as these will affect the price which he is prepared to pay for the property. In order to give maximum assistance to his client, in what will often be an important and difficult decision for him to make, the surveyor should provide a reasoned explanation of all defects and their probable consequences and also distinguish between serious defects and those of lesser consequence.

Intitial Procedures

Preliminary Procedures

Where a client wishes the assignment to be carried out on a fixed fee basis, the surveyor will need to undertake an exploratory inspection of the property in order to assess the time involvement and the nature and extent of any attendance required and/or specialist advice on services and equipment. In the latter case the client will need to be convinced that specialist advice is essential to obtain a comprehensive report on the condition of the property and all its component parts.

Attendance relates to the service provided by a builder, often charged on a daywork basis, which could include the provision, positioning and supporting of long extending ladders to permit inspection of otherwise inaccessible parts of the building, and the provision and use of drain testing equipment. Specialists include electrical engineers who inspect and report on electrical wiring and equipment, mechanical engineers who may be concerned with lifts, central heating and air conditioning systems, and structural engineers for structural aspects such as inadequate foundations and overloaded walls and structural frames.

The specialist will often require advice from the surveyor as to the nature and scope of his survey and report. As Bowyer[1] has described, an electrical engineer will normally be required not only to report on the type, condition and efficiency of an electric heating system but also its thermal adequacy, and such aspects as the adequacy of the main cable and fuseways where a substantial extension is contemplated. The report of a heating engineer usually incorporates the type of central heating system, energy source, adequacy and condition of the installation, and work required to bring it up to a suitable standard.

Initial Inspection

A RICS Practice Note[2] describes how a surveyor should familiarise himself with the district in which the property is situated and, in particular, with the character and nature of the property surrounding the building under inspection. The identity of the property must be checked on arrival, and with the client's instructions. The surveyor will make himself known to the occupier, where appropriate, or open up the premises where they are vacant.

The surveyor will note the principal design features and the extent and nature of the *accommodation*. It is not generally considered necessary to take detailed measurements at this stage, but the surveyor will make a note of any critical dimensions. Typical examples include pedestrian or vehicular side accesses of inadequate width; habitable rooms with inadequate lighting, ventilation, ceiling height or impractical proportions; staircases, corridors and door openings of a size or shape which restrict means of escape or movement of furniture. Any basements or cellars should be inspected and essential details noted.[2]

Extensive *site notes* should be recorded to provide a comprehensive picture of the property. The written particulars should incorporate the date of inspection, the names of all persons present, operative weather conditions, sources of relevant information and other significant factors. These particulars should be retained for as long as they may be needed. It is customary to ask the vendor or occupier questions concerning the history of the property as some of this information may be of considerable significance, and these details will be carefully recorded. This procedure will enable the surveyor to identify significant matters which may not come to light during the subsequent inspection.

The RICS Practice Note[2] lists a number of questions that might be asked and these provide a useful guide to the type of approach that could advantageously be followed.

(1) What is the age of the property and are there original or subsequent plans available?
(2) Has the vendor carried out any structural alterations or additions to the property and is he aware of any earlier modifications? If the answer to the last question is in the affirmative, the client should be consulted before the surveyor approaches the local authority to determine whether planning and building regulation approvals were obtained and the work carried out in accordance with the approved drawings.

(3) Have any structural repairs been carried out, including timber treatment, underpinning or strengthening, and do any guarantees exist?
(4) Has the property ever been flooded?
(5) Are there any concealed access hatches to voids, including those under floors, and is there a basement?
(6) Is there any dispute or claim with neighbours relating to boundaries, trees, access, drainage or other associated matters.
(7) Does the vendor propose to remove any items of equipment which would normally be regarded as fixtures, or to leave any items which are usually removed?
(8) How long has the vendor occupied the premises? The longer the period the greater will be the significance of the vendor's replies to many of the earlier questions.

Where the surveyor has engaged a builder to provide assistance during the survey, the surveyor should visit the site ahead of the builder in order to ensure that access to the premises is provided. He will also check the general layout of the drainage system and other services, locate the means of access to the roof space and determine the positions where assistance will be required in gaining access to those parts of the building which could be defective.[1]

Equipment Required

Principal Items of Equipment

The equipment required for a structural survey is influenced considerably by the approaches and procedures adopted by individual surveyors. Personal preferences based on long practical experience are bound to have their effect on the choice of equipment. The RICS Practice Note[2] includes a list of items of equipment which constitute the basic tools for this particular task and these are now listed, although it will be appreciated that alternative items may be substituted in some cases.

(1) A powerful, sturdy torch suitable for use in confined spaces and producing an even spot beam, preferably fitted with a strap for carrying purposes.
(2) Claw hammer and bolster for lifting and replacing floorboards, manhole covers and the like.
(3) Ladder with a minimum length of 3 m, which should be safe, foolproof, and easy to handle and transport.
(4) Pocket probe or penknife for initial testing of mortar and joinery, gauging fracture depths and similar activities.

(5) Binoculars, normally to a magnification not exceeding 8 for ease of operation.
(6) Hand mirror to a minimum size of 100 mm x 100 mm and often of shiny metal.
(7) Electrically operated moisture meter.
(8) Screwdrivers for removing electric covers and a variety of other tasks.
(9) Measuring rods or tapes, notebook and writing equipment.
(10) Plumb line for testing the vertical alignment of walls.
(11) Spirit level for checking the horizontal and vertical alignment of surfaces.

Staveley and Glover[3] include some other fundamental items in their list of equipment, such as Wellington boots, clip board, extension lead, electric socket adaptors, drain plug, protected lampholder for lead, spade, tool box and assorted tools and manhole keys. These are all based on years of practical experience of dealing with difficult conditions, when the absence of the required equipment can make the inspection of essential details very difficult. A protected lamp-holder, with a 100 or 150 W bulb, and an extension lead can be invaluable when inspecting roof spaces, areas below floors and other dark parts of a building, where a thorough examination of the structure and timbers is essential. A torch is often a poor substitute for mains-powered lighting.

Drain testing equipment should always form part of the survey equipment. In addition to drain plugs, manhole lifting keys, hammer and cold chisel and mirror, the author also believes in taking a pack of coloured dyes for the tracing of drains in doubtful situations. The spade can be used to take soil samples, expose hidden covers and even defective foundations on occasions.

Ladders and Access Plant

Hollis[4] distinguishes between the two main types of ladder — multi-section ladders and articulated ladders. Multi-section ladders are made up of a number of sections which slot together and are fixed with wing nuts. The four-section type with an extended length of 3.6 m and an unextended length of about 1 m is quite popular, as it will fit into a car boot. A set of steps often provides a better means of access into roof spaces. With articulated ladders, each section is connected by a locking hinge so that all sections of the ladder are joined together. This form of ladder can also be used to form a set of steps or staging, but it is heavier and occupies more space than the multi-section ladder.

With taller buildings, ladders are rarely adequate and more sophisticated equipment is needed. Mobile hydraulic platforms of the articulated boom or telescopic boom types will normally serve the purpose. The surveyor will need to check that there is sufficient space available for the plant with adequate headroom, a clear parking area and the ability to move the plant sufficiently close to the building for the horizontal arm to reach the desired position.

Measurement of Dampness

Dampness constitutes one of the worst problems in buildings and the surveyor must be able to measure with a reasonable degree of accuracy the moisture content of various materials, including timber, masonry and plaster. The measurement of moisture in building components can be undertaken in one of three ways — by a carbide moisture meter, by electrical resistance meter or by laboratory testing.

Carbide Moisture Meter

When using a carbide moisture meter, a small sample of walling material is placed in a meter and mixed with calcium carbide, when a gas will be released. The pressure created by the gas is converted into moisture content and can be read on the dial of the meter.

The hygroscopicity of the walling material can also be determined with the same equipment. The sample is divided into two equal portions — the first tested in the meter and the readings recorded and the second part placed in a 75 per cent humidity enclosure for about 2 days and then tested in the meter. The two sets of readings are compared to give the level of hygroscopicity, which demonstrates the relation of the moisture content to the maximum level of moisture that the material can absorb from the air. By this means it is possible to eliminate hygroscopicity as the cause of dampness.

As described by Hollis[4] the equipment has the disadvantage of being relatively heavy and bulky, and it takes 10 to 15 minutes to obtain the sample and the relevant meter readings. The taking of samples often results in damage to wall surfaces and decorations and the use of power drills requires a power source on the site. It does, however, produce accurate results.

Electronic Moisture Meter

This form of meter will measure accurately within certain ranges the moisture content of many species of timber and also provides an indication of the level of moisture present in other materials such as brick, stone and concrete.

Two prongs are attached to the body of the meter or to a cable used in conjunction with it. The two prongs are pressed into the surface of the timber and an electric current passes through one prong from a battery in the meter and is transmitted through any moisture present in the timber to the second prong. The meter measures the resistance to the passage of current between the two prongs. Water is a good conductor of electricity and thus the higher the amount of current passing through the meter the lower will be the level of resistance measured. The results are given as the percentage moisture content for timber and comparative readings for other materials. Meters have also been developed to measure the moisture content below roof surfaces and these are particularly useful in diagnosing failures in flat roofs, especially those covered with felt.

This form of meter has proved extremely popular, being small and light in weight, easy to operate and read, and causing minimal damage to decorations. It cannot, however, be used to measure hygroscopicity and results may be distorted with some materials, such as aluminium foil backed plasterboard and salts in old walls in the absence of a salt detector. In 1984, prices ranged from about £75 for pocket moisture meters to £460 for the more sophisticated digital instruments complete with all accessories.

Recent developments include the Protimeter Digital Diagnostic which can be programmed by means of plug-in program keys and is capable of being used with six accessories or attachments to carry out the following jobs.

(1) The measurement of moisture in timber from about 7 per cent to fibre saturation point near the surface or to a depth of 25 mm or more using a hammer electrode. Programs are available for over 150 species of timber and also for chipboard.

(2) The determination of the degree of moisture present in most building materials, including concrete, brick, mortar and plaster. The Protimeter has a colour code (green for safe, yellow for investigate and red for taking remedial action).

(3) The measurement of the level of moisture within walls, by drilling holes and using deep wall probes.

(4) The determination of the temperatures of wall, floor and ceiling surfaces and ambient air, using a wall thermometer with an integral program key. This attachment can also be used to determine ambient relative humidity and also the dew point using the wet and dry method.

(5) The determination of relative humidity and temperature, using a thermal-hygrometer attachment.

(6) The evaluation of contaminating salts in surfaces, using a salts detector.

Laboratory Testing

This procedure is used to measure accurately the moisture content and hygroscopicity levels of samples of materials such as brick, plaster, concrete and tiles. It is, however, expensive and the charge for a single test is equivalent to about half the cost of a small electronic moisture meter. Furthermore, damage is caused to surfaces in obtaining samples.[4]

Background to Moisture Measurement in Buildings

The percentage moisture content of a material is the amount of water in it divided by its weight, when dry:

$$\text{percentage moisture content} = \frac{\text{wet weight} - \text{dry weight}}{\text{dry weight}}$$

With the same amount of water, the greater the dry weight, the smaller will be the percentage moisture content. A heavy material has a much lower moisture content than a light material containing the same amount of water. For example at 3 per cent moisture content:

wood is dangerously dry;
most mortar is fairly damp;
most bricks are damp;
engineering bricks are wet; and
plaster is very wet.[5]

Conversely the following materials are air dry at the percentages stated:

wood – 15 per cent; lime mortar – 5 per cent; cement mortar – 2 per cent; some bricks – 2 per cent; other bricks – 1 per cent; and plaster at below 1 per cent.[5]

Materials under their generic headings are themselves infinitely variable in their composition. The weight of dry mortar will vary according to the ratio of sand and cement, while concrete will be affected by the different types of aggregates in addition to the ratios of the constituent parts. Clay for brick-making varies from region to region, while plaster can be one of a large number of mixes. Thus it can be seen that building materials are of infinite variety, and that determining the moisture content of a material, other than wood, does not indicate whether the material is wet or dry.[5]

Electronic moisture meters, such as those manufactured by Protimeter, give relative readings of free water in a material. Although they do not measure relative humidity, the readings provide a close representation of it. A high meter reading, in the absence of contaminating salts or carbonaceous materials, indicates a damp condition of approximately equal significance in wood, brick, plaster or wallboard, regardless of their varying moisture contents.[5] The best apparatus for measuring relative humidity is a hygrometer.

Electronic moisture meters measure wood moisture in wood and wood moisture equivalent in building materials other than wood. Hence it is possible to incorporate on the scale of a Protimeter areas designated safe, intermediate and danger, which correspond reasonably well with the humidity equilibria of most non-metallic or non-carbonaceous materials on which they may be used. This is done by a colour code wherein green indicates a safe condition, corresponding to an air-dry condition in a normal indoor, inhabited environment, red indicates a humidity equilibrium in excess of 85 per cent, and a yellow area indicates the area in between these extremes.[5]

Materials do not become visibly damp nor feel damp to the touch up to 85 per cent humidity equilibrium. For example, wood does not feel damp below a moisture content of 30 per cent (around 97 or 98 per cent relative humidity), although rot will develop at 20 per cent moisture content and above. Hence, dampness reaches dangerous proportions long before it can be detected by the unaided senses, and this is why the electronic moisture meter is such a valuable item of building surveying equipment.[5]

Summing up, if an absorbent material is placed in a very damp atmosphere, its moisture content will change until it equals the humidity of the surrounding air. If however the material was saturated prior to placing in a damp atmosphere, its moisture content could fall or, if it was dry when placed in the damp atmosphere, its moisture content would rise. Variations in the dampness of the air around an absorbent material, or fluctuations in the relative humidity of the air, will produce changes in the moisture content of the material. As described earlier, the moisture content of different materials will not be the same when they attain a point of equilibrium with the same relative humidity in the surrounding air. The growth of moulds and decay fungi require a high relative humidity, in the order of 75 to 85 per cent, when the timber is likely to have a moisture content between 17 and 20 per cent.[4]

Sequence of Survey Operations

General Sequence

It is advisable to adopt a logical sequence when surveying a property to minimise the risk of omissions, to keep the time spent to reasonable proportions and to enable the report to follow logically from the inspection.

A satisfactory approach to the survey of a domestic building would be as follows:

(1) General description of building and site.
(2) Roof voids.
(3) Room by room inspection.
(4) Staircases.
(5) Any accommodation below ground floor.
(6) Sanitary appliances and services.
(7) External elevations.
(8) Drainage.
(9) External works and adjoining properties.

Detailed Procedure

Each of these sections will now be examined in some detail to illustrate the scope of the matters to be investigated and noted. All parts of the property should be surveyed as far as practicable and all access hatches and panels opened for this purpose.

General Description of Building and Site

The surveyor usually starts by noting the general character and nature of the property, including the type of accommodation, such as detached, semi-detached, position within a terraced block, house with number of storeys, flat, bungalow or converted building. In the case of a semi-detached house it is customary to indicate whether it is left or right handed, and with terraces the length and number of dwellings in the block are worthy of mention.

The surveyor will be concerned with the general character, quality, age and condition of the property. It could be a former local authority dwelling sold to the occupier, a house on a speculative development built in keen competition, or a one-off house which was architect designed and built to a client's specific requirements. The age of the property is always important as it is likely to influence the extent of the defects and the amount and cost of remedial work to be undertaken by the purchaser. A property less than ten years old is likely to be the subject of a National House-Building Council warranty, while a dwelling which is 30 to 40 years old may require some major works to be undertaken, such as retiling or slating, pointing of brickwork, rewiring of the electrical installation and plumbing repairs and replacements. The age can be ascertained by inspection and enquiry.

The general condition of the property must be recorded and this can vary from a very well maintained dwelling to a grossly dilapidated and neglected building. Missing and flaking roof tiles, crazed and missing rendering, cracked brickwork, rotting window sills and neglected paintwork indicate the need for an extensive investigation and the listing of numerous defects.

The area and shape of the site, degree of privacy, nature of boundaries, type of soil and groundwater level, extent of falls, type, size and location of trees all need carefully recording, and any of these aspects can be of particular significance. A combination of shrinkable clay subsoil and large trees growing near the building can give rise to serious settlement problems. Record details of outbuildings, including in particular the capacity of the garage(s) or any buildings suitable for conversion into garages. The type and condition of the drive and paths and scope and character of the garden should be examined. It is important to establish that the dwelling is erected on firm ground, and buildings constructed on made-up ground need further investigation to determine the nature of the fill and likely degree of consolidation. A study of geological maps and suitable enquiries will assist in establishing the facts.

Roof Voids

Access to the roof voids is essential to establish the type, size, span and condition of roof members. A sketch of the roof members with a note of their size and spans will be useful. A note will also be made of any roof insulation, details of services, including the water storage tank, party and gable walls, fixing of roof coverings, type and condition of chimney stacks and flues, types of ceilings, area of boarding and all other relevant details. This is an important part of the structure and deserves careful scrutiny, even if the going is rough when balancing on ceiling joists and climbing between roof members. A cautious and steady approach, with the provision of ample lighting, is essential.

Room by Room Inspection

The first step is to draw a sketch of each floor plan, working from the top of the dwelling downwards. The rooms must be of the correct proportions and some surveyors use squared paper and draw the plans roughly to scale. Dimensions are inserted in the manner shown in the next section of this chapter. Each room will be described or numbered for ease of reference.

With regard to the individual rooms, it is customary to start with the floors, noting the type of floor covering and its condition. Problems may arise with fitted carpets and consent should be obtained to lift sections to enable a full examination of the floor coverings to be made. Boarded floors can be tested by walking along them and springing on the centre of the spans. Where unsound floor construction is suspected, floorboards should be taken up and the joists examined for adequacy of support, size and condition. Defective joists will need replacing and attention should be given to the work required to the ceiling below. Ground floors will need checking for possible infestation and rising damp.

Walls should be checked for any defective plaster by tapping the suspected area; if unsound, a hollow note will generally be emitted. The type and condition of ceiling and wall decorations, cornices, coves, picture rails, chair rails and skirtings will all be noted, and a standard format could assist in the entry of this type of information, with suitably headed columns being incorporated.

The surveyor will follow by entering detailed of windows, doors and other joinery and also fireplaces, radiators, lighting and power points and any other fittings and services. When recording these details, he will also have regard to their adequacy and the extent of the work required in making good any defects, in order that this can subsequently be entered on a priced schedule.

Staircases

It is probably the best practice to examine all staircases together, rather than dealing with them separately on a floor to floor basis. The type of construction, dimensions, adequacy and condition all require noting.

Accommodation below Ground Floor

With older dwellings it is quite common to find basements and cellars, which may be devoid of proper damp-proofing treatment. In consequence expensive remedial works may be needed to render them usable. In recommending water-proofing work, regard must be paid to the probable life and value of the property. Whereas asphalt tanking might be the ideal solution, the age and condition of the property might not justify such expensive treatment. A cheaper solution, such as the application of synthaprufe, is likely to be a more realistic proposal.

The form and condition of the construction, form of damp-proofing if any, and means of ventilation will all require careful scrutiny. The surveyor will look particularly for signs of dampness and the cause of penetration, in order to decide on the scope and nature of the remedial work involved. Any timber will be examined for possible insect or fungi attack in these very vulnerable areas. Structural instability may constitute a problem in some cases. All in all, basements can constitute one of the most troublesome parts of the building.

Sanitary Appliances and Services

All sanitary appliances should be carefully inspected, including associated taps, traps, water waste preventors and valves, and tested by passing water through them and looking for leaks. Cupboards and timber floors adjoining sanitary appliances should be examined for any signs of rot or decay.

The RICS Practice Note[2] describes how hot water installations, boilers and control equipment should be activated to test physical operation, and inspected visually for any signs of corrosion or leakage. The type and condition of space heating appliances should be noted and checked in a similar manner to the hot water installation.

Gas installations should be checked visually where exposed and the gas appliances inspected and operated if the gas supply is connected. Electrical installations should be checked visually to the extent required to advise on the probable age and suitability of the wiring or equipment and its likely condition. The surveyor will not normally carry out insulation tests but he should give an opinion on the need for such tests.[2]

The connected main services should be described and the availability of other services established. The surveyor normally advises the client that further tests by appropriate specialists will be required if he wishes to receive assurances as to the condition or capability of the services.

External Elevations

Some surveyors prepare sketches of each elevation of the building on which any significant aspects are suitably recorded. For example, with fractured walls, the fracture lines should be drawn on the sketches, showing the starting and finishing points and the widths at suitable intervals to give a clear picture of the nature and extent of the defects. The walls should be carefully inspected to identify and note any defects arising from settlement, bulging walls, defective pointing or rendering and the existence or otherwise of effective damp-proof courses. Copings and parapet walls subject to exposed conditions require careful scrutiny.

Marks or stains on walls often result from leaking overflow pipes, gutters and downpipes, and their positions should be indicated on the sketches. These positions should be checked internally to determine whether there is damp penetration in identical locations.

The inspection of elevations should also encompass exterior joinery, particularly lower window and door frames, fascias, soffit boarding and barge boards. Other vulnerable areas include balconies, porches and bay windows, and these should receive thorough examination. All elevations should be carefully checked for any signs of structural movement, using a plumb line and spirit level where necessary. Evidence of remedial work in the past should alert the surveyor to possible problem areas.

Brickwork within accessible construction voids should be inspected as far as practicable and any opportunity taken to examine flues and wall cavities. All vulnerable areas of walls should be tested with a moisture meter. Where structural defects indicate the likelihood of problems below ground, wall foundations should be exposed by the surveyor or a contractor after obtaining approval from the client.

All roof areas should be inspected as closely as possible using all available equipment and vantage points. Particular attention should be paid to missing, slipped, cracked or laminated tiles or slates, or any other signs of deterioration of the roof coverings, supporting nails or battens. The bedding mortar to ridge and hip tiles is frequently in a soft or loose condition and thus requires attention. Valleys should be inspected and secret gutters, in particular, are vulnerable to blockage by dead leaves and other debris.

Chimney stacks require careful scrutiny as they may have cracked or loose pots, defective flaunching, cracked brickwork, and unsound flashings, soakers or cement fillets at the junction with the roof covering, often resulting from movement of the roof timbers, which can permit rainwater to gain access into the roof space down the sides of chimney stacks. The treatment of redundant flues needs observing and noting.[3]

Flat roofs are a common source of water penetration and justify special attention. The blistering or cracking of both asphalt and bitumen felt are frequent causes of defective flat roofs, coupled with inadequate falls and defective skirtings or upstands. Where a defective covering is encountered, a thorough inspection of supporting boarding and roof timbers is required, as a leaking covering could have resulted in extensive timber decay.

A close inspection should be made of all rainwater goods, paying particular attention to the insides of gutters and adequacy of brackets, condition of downpipes and all associated rainwater fittings. Leaking rainwater pipes and gutters and overflowing heads can, if left unattended over a period of time, result in damp penetration through walls.

Drainage

All accessible manhole covers should be raised and the manholes examined and the flow through the manholes checked. An inspection of the drains between manholes should be checked with a mirror. Where manholes are silted up or in poor condition and drainpipes cracked, partially blocked or unsatisfactory in other respects, this should be reported to the client. The surveyor may carry out drain tests after arranging for any blocked pipes to be cleared, or recommend that such tests be undertaken.

The lines of the drains should be traced and recorded, distinguishing between foul, surface water and combined drains. The effectiveness of the drainage system and the disposal arrangements should be assessed and included in the report. At the very least all gullies and drain runs should be the subject of a minimal test by observing the flow of water through the system. Above ground drainage pipes should be inspected and preferably be subjected to a smoke or other appropriate test. If left untested, this should be stated in the report. It may also be necessary to establish the ownership of drains, as some of them may be in joint ownerships.

External Works and Adjoining Buildings

The inspection of fences, boundary walls, outbuildings, paths and other features on the site should be undertaken to the extent necessary to report on matters included in the client's instructions.[2] The client may consider that it is not worth while to pay the surveyor to inspect and report on fences, gates, sheds, greenhouses and associated structures and fixtures. However, there can be instances where the outbuildings and boundary walls are such extensive structures that they justify a full survey and report, particularly with larger, older houses. For example, there may be large stable blocks converted into garages and massive stone retaining walls supporting parts of the site, where extensive repairs could be very costly. Hence each case needs separate consideration, having regard to the particular circumstances.

The ownership and liabilities in respect of boundaries need clarifying, having regard to the title plan and covenants. A 'T' is often printed against a boundary, indicating that responsibility for maintenance rests with the owner of the land on which the 'T' appears. Should the verbal description of the parcels and the plan disagree, the construction of the deed's reference to the plan will normally determine which will prevail.[6] In this type of situation legal advice should be sought.

Walls which are shared by two properties, such as between semi-detached or terraced houses, may belong to one of the adjoining houses exclusively if there is clear evidence that the wall is on one side of the boundary line. It is, however, more usual for these to be party walls, and in the absence of contrary evidence there is a presumption that the boundary runs down the middle of the wall.[6]

The general condition, construction, design and use of adjoining properties should be observed and recorded, in order to identify any particular factors which may adversely affect the property being surveyed.[2] The surveyor should also check for any possible future development or redevelopment of neighbouring sites and any proposed highway improvement schemes which could affect the property. Other relevant factors include fronting on to unmade or private roads, existence of building lines, and restrictive covenants and easements limiting the prospective owner's rights over the property. These matters can have a significant affect on the value of the property and need further investigation by the client's solicitor.

Recording of Site Notes

Use of Check Lists

Check lists are often used as a basis for the preparation of site notes when carrying out surveys of buildings. They are normally subdivided into sections or elements for ease of inspection and to provide a logical sequence when writing the report. Bowyer[1] describes the two principal methods of using check lists.

(1) A comprehensive check list is completed by the surveyor as he inspects the building and this reduces the risk of omissions and results in the information being provided in a standardised form.
(2) A check list provides a useful set of headings which the surveyor can use as a basis for dictating the report to a secretary.

Specimen survey check lists are illustrated in appendix A and appendix B, covering site factors and building elements respectively.

Drafting of Reports

The drafting of the report can be done by hand, by dictating to a secretary or by dictating the report on to tape either in the office or on site. Dictating on to tape should be the most economical method, but some persistance may be required before the surveyor becomes proficient. Views vary as to whether it is better to dictate the report on site or in the office. Some surveyors contend that dictation on site saves time and permits greater flexibility in marshalling the relevant information while inspecting parts of the building, without the distraction of writing. Writing is made more difficult in the rain and in dark, cramped roof spaces. Objectors to this method argue that the portable tape-recorder can malfunction, does not readily back-track and cannot be used with confidentiality on site.

Surveyors are under pressure to produce complex structural reports quickly and economically and Braine[7] believes that the use of a word processor is one method of achieving this. Some surveyors use word processing equipment for text editing to prevent extensive retyping. The more sophisticated the equipment, the less likely it is to be economical, although there are staff savings.

Problems Associated with Building Surveys

General Problems

Surveyors carrying out structural surveys face various difficulties. Often they can see only what is present on the surface, and they have to rely on their professional skill, experience and intuition to deduce the significance of the symptoms that they observe. Even in the most favourable situations, the surveyor is usually permitted only to take up a few floorboards, enter the roof space, probe timbers provided he does not damage finishes and decorations, lift manhole covers and test drains, electrical installations and other services.[8]

The surveyor cannot normally cut holes in ceilings to gain access to roof voids where there are no access hatches, and rarely can he take up sufficient floorboarding to examine the joists and plates below or the condition of the concrete oversite, mainly because of the problems involved in making good the resultant damage. He is usually unable to investigate foundations without the client's permission, which is likely to be withheld if it results in breaking up pavings or damaging flower beds. Even when he is given the opportunity to inspect deposited plans at the local authority's offices, there is no guarantee that the building was erected strictly in accordance with the submitted and approved plans.

Where floors are covered with fitted carpets and rooms are almost completely filled with items of large and heavy furniture, the task of the surveyor is made very difficult, unless he is assisted by expert furniture removers and carpetlayers. Most vendors are likely to object to the entry of so many people on to the premises and the prospective purchaser may not be prepared to meet the additional costs involved.

A structural survey has been described as primarily an exercise in intelligent guesswork, carried out by someone who is able to interpret the significance of visible features. However, the value of structural surveys must depend to a large extent on the skill, experience and diligence of the surveyor undertaking all possible investigations on the site and painstakingly diagnosing all conceivable faults and their likely consequences. These must all be carefully and meticulously documented in the report, so that the client is made fully aware of the condition of the property, its shortcomings, the extent of the remedial work required and the costs involved, and their relative priorities.

The needs of the client must be clearly identified and satisfied. He could, for instance, require a detailed examination of the structural condition of the building in all its aspects, including the adequacy of foundations, the load-bearing capacity of floors and walls, and the practicability of adapting the property to an entirely different use, such as the conversion of a hotel into a private nursing home or a church into a car showroom.

Timber Frame Houses

A surveyor is generally not instructed to open up parts of the structure for examination but usually carries out certain diagnostic actions. Problems could therefore arise if the surveyor was unaware that he was surveying a timber frame house. Melville[9] has described various techniques of identification.

Windows are probably the best indicator, as they are usually fixed to the inner timber frame, resulting in a deeper than normal reveal. There is often differential movement around the window opening and so a 5 mm gap is usually allowed for shrinkage in the structure which is likely to take place in the first two years after construction. Where brick cladding is provided, the overall wall thickness will be greater in a timber-framed than a traditional house. Weepholes will generally be provided in external walls below damp-proof course level. Melville[9] suggests the use of a compass or magnet to locate nails on internal walls, indicating the location of studwork, which can then be tested for dampness.

Another useful approach is to remove a switch cover from an external wall or to look through a floor trap. If the party wall in the roof space is plasterboarded, this indicates timber frame construction, and blockwork need not necessarily mean that the structure is traditional. The top of a wall in the roof space could reveal the type of construction, and rafters in a timber frame house are usually planed whereas those in a traditional house are sawn.

Subsidence

General considerations A surveyor carrying out a structural survey must be able to diagnose the significance of cracks in walls. He must, for example, be able to recognise a crack that is old and inactive and that the circumstances which caused it no longer operate.

What is required in a structural survey report is a factual statement of horizontal levels, perpendicularity and crack patterns and sizes. *BRE Digest 251*[10] sets out clearly a form of reporting which should ideally become the standard approach. The report consists of four components:

(1) Elevational sketches showing crack patterns and widths, both internal and external.

(2) Diagrammatic record of horizontal levels, which can be assessed at a brick course, plinth top or any other horizontal feature in the elevations.

(3) Diagrammatic record of verticals at strategic points around the structure.

(4) A commentary on any aspects of structural movement not clearly shown by the three foregoing items.

Assessing the prospects of future structural movement is more difficult, but there are certain factors on which a reasoned conclusion can be drawn. The first step is to eliminate damage caused by leaking drains or water services. Then the foundation construction and subsoil conditions should be investigated. If the surveyor is outside his own locality, he should make enquiries about local land-slip, made-up ground and mining subsidence. Local knowledge is sometimes more valuable than exhaustive investigations.

The existence of mature trees close to a house is usually regarded as a hazard but their removal, in some circumstances, can be worse. Hence it is important to determine whether the house preceded the tree or vice versa. In the case of new properties, particularly in areas of shrinkable clay, it is advisable to find out whether the site was wooded before development. As a result of these investigations it will be possible to reach a conclusion as to the possibility of further settlement.

Financial aspects House owner insurance policies now provide cover for damage caused by subsidence or landslip of the site, subject to the limitations of the policy wording, which is often subject to exceptions and an excess which applies in respect of each and every loss.[11]

In the case of *Allen (David) & Sons Billposting Ltd v. Drysdale* (1939) 4ALL ER 113, Lewis J. made a comment that 'subsidence' means sinking, whereas 'settlement' means movement sideways. Insurers will readily accept the downward movement of the site as subsidence within the meaning of their policies. The next consideration is whether the movement is of the site on which the house stands. Almost all insurers use the wording 'subsidence of the site'. The site is normally regarded as the virgin material beneath the foundations or under the floors of a building, while fill or hardcore which was placed in position when the building was erected is not regarded as part of the site.[11]

Landslip Landslip is usually regarded as a sudden massive movement of land following structural failure of a bank or retaining wall. There is normally an exclusion covering boundary walls and if these suffer landslip in isolation, the cost of rebuilding will not be covered by the owner's policy.

Ground heave Insurers do not always provide cover against ground heave. A typical example is where houses are built on a clay subsoil across the line of a mature hedgerow or tree which was cleared in connection with the development. After completion of the houses, the clay subsoil, which was kept relatively dry by the hedgerow or tree, takes up moisture; the clay swells, causing the building to rise with consequent cracking of the superstructure. Levelling and monitoring of the movement will identify the cause of the problem.[11]

New Buildings

The survey of new buildings, where they encompass new materials and untried techniques, can prove more difficult than that of older buildings. In contrast, older buildings generally produce fewer problems of diagnosis since they normally incorporate well-tried materials and methods. Where old buildings are neglected, they can usually be relatively easily repaired and restored.

Some materials used in new buildings, such as high alumina cement and woodwool slabs, have a relatively high incidence of failure. System-built houses have the highest danger rate and BRE reports[12] confirm that defects to prefabricated reinforced concrete houses are caused by corrosion of reinforcement through carbonation, and in some cases through high chlorine levels leading to acidification of the concrete.

Defects such as serious corrosion of reinforcement and subsequent spalling of concrete, effects of excessive condensation and rapid deterioration of poor-quality joinery are readily visible, but fixings for claddings and the presence of unexpected materials which are likely to be harmful, such as blue asbestos and urea formaldehyde foam, make identification difficult. With some new building types and forms of construction, there has been only restricted feedback of maintenance data, and so there is greater difficulty in determining the likelihood of faults than with traditional buildings.

Adaptation of Redundant Buildings

Documents and Drawings

Eley[13] has identified the sources of background information which can help the surveyor when he is conducting a survey of a redundant building, which it is proposed to use for another purpose.

The main sources are as follows:

(1) The present building owner and his agents, such as the surveyor and solicitor.
(2) The local building control authority.
(3) Insurance plans covering the commercial and industrial areas of most cities and towns.
(4) Local libraries and record office: these may contain old directories of the area, from which earlier owners and change of use can be determined, and possibly a tentative date for construction, and other information such as

local histories, large scale Ordnance Survey maps, industrial archaeological surveys and reports.

(5) Original and later architects and engineers, as most practices retain their original drawings or microfilm copies. The RIBA Drawings Collection and the Institution of Civil Engineer's library could also be approached, particularly if the designer were well known.

(6) Architectural, building and engineering periodicals where a construction date is known with reasonable accuracy; *The Builder, The British Architect* and *Proceedings of the Institution of Civil Engineers* may be able to provide an informative account of the then newly erected building.

(7) Archives: The Historical Manuscripts Commission publish lists of sources known to the National Register of Archives, and they include business and architectural history.

(8) Geological Survey maps and memoirs provide excellent sources for information on ground conditions. The local authority may also be able to supply information and identify site investigations carried out on nearby sites.

(9) Public authority buildings: where a building is now or was once in the possession of a nationalised industry or public authority, an approach to the appropriate body could be fruitful. Typical examples are:

Canal buildings: British Waterways Board

Railway buildings: Public Record Office for British Transport historical records and the Regional Civil Engineers, London Regional Transport Authority

London dockside buildings: Port of London Authority

Colliery buildings: National Coal Board

Steelworks: British Steel Corporation.

Survey of Existing Structures

The cost of structural work required to adapt a redundant building to new user needs and to ensure its compliance with statutory requirements can often be the major part of the cost of a project and so determine its feasibility. It is therefore necessary to establish at an early stage the scope of the work required, and Bussell[14] describes how this can be achieved by surveying the existing structure, assessing the structure as it stands and then assessing it for its new use. The survey needs to be very exhaustive and is often undertaken in two phases – an initial survey followed by a detailed survey.

Initial survey An initial survey is undertaken to obtain a general picture of the overall condition, materials, construction and major defects. Time is often restricted but it must be sufficient to enable any significant problems to be identified which could affect the decision to proceed further. Survey data can be obtained from any available documents and drawings, supplemented by an examination of the structure itself. It is also advisable at this stage to consult the local building control authority to establish the extent of the structural information required in a submission under Building Regulations.

Detailed survey A detailed structural survey will subsequently be undertaken to obtain all relevant structural information. It will often be necessary to remove finishes at selected points to establish the size and other details of structural components. The extent of the investigation will, however, be influenced by the available documentary evidence, the degree of standardisation of structural components, the designer's needs and the impression gained on the general inspection of the building and its visible defects.

Traditional buildings in masonry and timber can generally be surveyed using the procedures applicable to house surveys, as described earlier in this chapter. Construction in metal and reinforced concrete often requires advice from a structural engineer, as many specialist aspects are involved as described in the following paragraphs.

The assessment of the *in situ* strength of structural members may involve sampling. Timber species can be identified by TRADA (Timber Research and Development Association) or the BRE Forest Products Research Laboratory if required. Bricks and mortar can be taken from lightly stressed areas, to be crushed and analysed in a materials testing laboratory. Concrete can be cored and reinforcement sampled from a lightly loaded area for tensile testing. Iron or steel members are normally confined to visual inspections but they can be sampled with care for analysis and strength tests if required.

Concrete slabs and walls can be drilled to determine the thickness more accurately than by calculating it from overall dimensions. Rolled iron and steel sections should have web and flange thicknesses recorded. The nature and thickness of applied finishes are also usually recorded.

The surveyor or engineer conducting the survey will look for signs of distress in the structure and record such items as sagging floors and roofs, bulging or out-of-plumb walls, pattern and widths of cracks, distortion of members and movement at joints. Cracks require special attention and it needs establishing whether they are static or live, recent or old, and significant or minor. Telltales can be placed across cracks to determine whether movement is continuing and should preferably be steel studs, the length of which can be monitored by a strain gauge, unlike glass strips.

Bussell[14] also recommends the checking of joist, slab and beam bearings, particularly where there are signs of movement such as walls out of plumb and spalled finishes under bearings. Inaccessible connections in reinforced concrete and encased steelwork may have to be exposed or X-rayed, although both can be hazardous and expensive. The surveyor or engineer should measure and sketch uncased connections in metal-framed structures, noting plate thicknesses and the sizes of bolts, rivets and welds.

When inspecting substructural work, trial holes will normally suffice for shallow foundations, but where deep foundations are likely, every effort should be made to find the relevant documents. Foundation material, size and condition should be noted. The bearing strata should be inspected, probed and possibly sampled. A thorough site investigation involving deep boreholes may be necessary, but this will be expensive. Trial holes should be carefully sited, adequately supported while open and backfilled as soon as possible after inspection.[14]

Surveys of Flats

Lease and Service Charges

The surveyor when carrying out the survey of a flat has to give his advice on the condition of the individual flat set against the problems in the main building. The owner of the flat will almost certainly have to pay a proportion of the outgoings on the main building, although it could be a relatively small percentage. A detailed inspection of the complete building is necessary before the surveyor can advise on the extent of the flat owner's liability.

The surveyor should comment on the pattern of past expenditure on the building and advise on the probable trends of future service charge expenditure, based on the likely maintenance and replacement costs of such elements as lifts, communal heating and hot water supply, roof, walls and exterior decorations in the years ahead. Ideally, the surveyor should inspect the accounts for the maintenance of the building over the previous three to four years.

The surveyor should draw the client's attention to any specific clauses in the lease which could interfere with his enjoyment of the flat. The effect of the clauses covering alterations and covenants should be explained. Many leases now require flats to be carpeted throughout and this can be a costly item.

The Survey

Hollis[4] has described how the main expenditure in the operation and maintenance of flats falls into the following categories – roof; exterior decorations; lift maintenance; communal heating and hot water supply; staff and staff accommodation; and interior decorations. Access is required to concealed parts of the roof, lift room, communal boiler and central heating plant, so that their condition can be included in the report.

The detailed examination of the lift and associated equipment should be undertaken by a specialist engineer and it is advisable to obtain a copy of the latest engineer's report from the landlord. However, the surveyor can make general comments on the capacity, adequacy and condition of the lift(s).

Where the building has a communal heating and hot water system to serve the flats and common parts, the surveyor should report on the type of system and form of energy used, and the materials and condition of pipework, valves, tanks and cylinders. The caretaker's flat should also be inspected to obtain an indication of probable future maintenance costs.

Means of escape in case of fire must be examined to determine their compliance or otherwise with statutory requirements and any relevant fire certificate, and the general condition and the probable cost of any necessary works estimated. At the same time the standard of fire resistance of doors should be noted and reported.

Surveys of Commercial and Industrial Buildings

Initial Preparations

The surveyor should start by obtaining copies of all available building plans, agency details, Ordnance Survey and Geological Survey maps and other documents listed earlier under 'Adaptation of Redundant Buildings'.

Some authoritative guidelines are contained in a RICS Guidance Note[15] on which much of this section has been based. For instance, the surveyor must be fully cognisant of the extent of the property to be surveyed and the tenure of its component parts. It is an advantage if plans attached to title deeds and leases can be provided. The surveyor should inspect any lease documents prior to his inspection. He has a responsibility, in liaison with the client's solicitor, to advise on the effects of repairing covenants, any schedules of condition, rebuilding clauses and provision for reinstatement in the event of an insurable loss and any other practical implications.

Where a prospective purchaser takes over a liability to contribute towards the cost of maintenance and repairs undertaken by others, the surveyor must inspect all accessible parts of the building(s) covered by the service charge. It would, however, be advisable to qualify this section of the report as the inspection may be restricted.

It is often necessary for a surveyor to have a letter of authority as access may be subject to safety and security checks. He should also carry means of identification incorporating a photograph.

The surveyor should at this stage advise on the employment of specialists for specific investigations, such as structural and services engineers. They will be appointed as agents for the client but the surveyor will co-ordinate their work, including their reports. The surveyor should also determine whether he requires any specialist equipment for the survey, such as platform hoists. He will also need to check that he is covered for insurance against damage by the equipment to property.

The Inspection

The surveyor will inspect all accessible parts of the building(s), as the report will describe the form of construction of the various elements or components and draw attention to defects in design and construction and diagnose the causes and recommend remedies. The surveyor will on occasions suggest the need for further investigations. Minor defects which are readily identifiable by the lay observer are not normally itemised unless they are symptomatic of more serious faults. The survey will extend to boundary walls and fences, pavings and other external works.

The RICS Guidance Note[15] identifies the main considerations with regard to such structural elements as:

(1) strength and stability;
(2) dimensional stability;
(3) exclusion of water;
(4) ventilation of the structure;
(5) sound and thermal insulation;
(6) daylighting; and
(7) durability, composition and maintenance, including the assessment of probable future costs.

Services frequently form a major component in these classes of buildings. They can encompass electrical heating, hot and cold water supplies, air conditioning, lifts, hoists, escalators, moving platforms and drainage systems. Additional services may also include fire alarms, intruder alarms, telephone and communications systems and window-cleaning equipment. The surveyor's role is normally to carry out a general inspection of the services and to decide the type of specialist tests required, and subsequently to co-ordinate the specialists' reports for presentation to the client. Where alterations or additions to the buildings are contemplated, the practicability of extending the existing services requires investigation.

The surveyor will critically examine the energy conservation elements in the construction and advise the client on any deficiencies in thermal insulation arrangements and suggest economical methods of remedying the shortcomings. He should also warn the client of condensation problems and of the consequent damage which could be caused to the fabric, plant and furniture.

The surveyor must have in mind the client's proposed use of the building and any special requirements, and will advise as to the suitability of the property for the client's purposes and the need for and cost of any necessary alterations. He will need to consider the effect of general and building legislation, health and safety, fire precautions, disabled persons' needs and public health requirements.

The surveyor must inspect all available fire certificates and statutory approvals, particularly when alterations have been carried out, and note any changes which may not comply with the fire safety regulations. Where the documents are known to exist but are not made available, this should be recorded. Fire precautions fall into three broad classifications – fire separation, means of escape and fire protection. The surveyor will also be on the lookout for any shortcomings in security arrangements against unlawful entry.

The surveyor should consider carefully the possible effects of adjoining land and buildings on the property being surveyed, with particular reference to party walls and fences, areas in common use, easements and wayleaves (confirmed or suspected), potential nuisance and incompatible uses. In addition he should consider the environmental and physical factors which may affect the building and its proposed use, such as flooding, mining and other subsidence, trees, noise and obnoxious fumes.

Measurement of Buildings

Measurements for Property Particulars

A joint RICS/ISVA working party produced a code of measuring practice[16] to ensure a minimum standard of accuracy and uniformity of practice in estate agents' property particulars. It will probably be helpful to the reader to give the more important guidelines.

(1) Measurements should be taken to internal wall finishes and to the backs of fitted cupboards.
(2) Chimney breasts, projections and bay window recesses should be ignored, unless they occupy more than two-thirds of the length of the wall on which they stand.
(3) Where measurements are taken into bay or bow windows, these should be taken into the greatest depth and followed by such words as 'into bay'. Conversely, where window recesses are ignored, it would seem appropriate to state 'excluding bay'.
(4) 'L' shaped rooms are to be measured in two separate parts.
(5) Measurements of hallways and bathrooms should not normally be quoted unless of unusual size and, in the case of hallways, it should be made clear as to whether any dimensions quoted are inclusive or exclusive of the staircase.
(6) Within kitchens, measurements should normally be taken to internal wall finishes above worktops, and specific reference made to the existence or otherwise of fitted cabinets and the like, occupying usable floor space.
(7) Garages should be measured overall internally between the main wall faces, ignoring projecting piers and door reveals unless these are unusually restrictive, when the effective minimum width should also be stated.

(8) The longer of the two measurements in any room should be quoted before the shorter.

(9) Property particulars should state that all measurements given are for the purposes of description rather than for calculating areas, and that they should not be relied on for other purposes such as the ordering of fitted carpets. Furthermore, that reliance should not be given to any plans supplied.

Measurements of Buildings to Produce Scale Drawings

As an entirely separate and distinct operation from the measurement of buildings to assist with the preparation of property particulars, it is often necessary to measure existing buildings in detail to prepare plans, elevations and possibly sections, for record purposes or to provide a basis for the design of alteration work. The main essentials are accuracy, completeness and neatness in preparation. There is nothing more annoying than to have to return to the site to take one or two additional measurements which had either been overlooked or are subsequently found necessary in preparing details of alterations.

Some surveyors prefer to use a field book for recording measurements on site, which can either be unlined or made up of graph paper. This approach obviates the possible loss of sheets but the page size is restricting and can result in cramped sketches. A better alternative is to use loose sheets of paper on a clipboard, each headed with the name of the building. The use of graph paper will help to keep the plan rectangular or alternatively will highlight any divergences from the rectangular shape. It will also be easier to keep the various parts of the building in approximately correct proportions. McDowall[17] has described how lengths can be assessed in terms of the width of a doorway, for example lengths of wall or widths of windows may be judged as equal to one, two or four-and-a-half doorways as appropriate. Starting with a suitable number of squares on the graph paper for a door width, other dimensions can then be assessed in terms of the number of squares on the paper and the whole plan kept in correct proportions.

The surveyor should start work on the site by inspecting the whole building to become familiar with its layout on all floors, and to note variations between the positions of internal walls and partitions on different floors. He will then proceed to sketch a complete floor plan usually starting at the top of the building and continuing downwards.

The instruments used for measuring buildings are a linen tape, usually 20 m long, which is easier to handle than steel and sufficiently accurate for most surveys, but requires two persons, a folding wooden measuring rod often 1.5 m in total length, and a steel pocket rule usually 2 m in length. The wooden rod is useful for measuring relatively short lengths where the ends cannot easily be reached, and the steel rule for measuring short distances between projections.

Ideally each wall should be measured in one series of running measurements, all taken from one end of the wall. For example, if the side of a room is 3 m long and contains a window 1.050 wide, a continuous set of measurements of 750, 1.800 and 3.000 is preferable to individual dimensions of 750, 1.050 and 1.200. The running measurement approach is likely to be more accurate and saves time, but can be carried out only with a tape. Typical examples of both methods are illustrated in figure 1. The method of recording dimensions varies but one approach with running dimensions is to insert an arrowhead where the zero end of the tape is held, a small dash at right angles to the wall at each point to be measured, and to enter measurements at each dash at right angles to the length of the wall. Overall lengths are entered parallel to the wall face.[17]

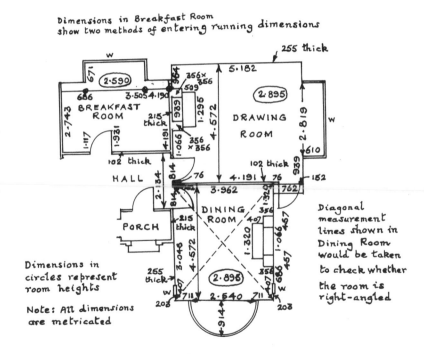

Figure 1. Measuring buildings for floor plans

In addition to taking overall dimensions of rooms and widths of door and window openings, it will also be necessary to measure diagonals as the room may not have right-angled corners. All wall and partition thicknesses must be noted and the room heights are normally inserted in circles for ease of identification. It is important that the sketch plans are drawn sufficiently large to prevent the details and dimensions from becoming cramped and difficult to read at the plotting stage when scale drawings will be prepared. It may not always be considered practicable to take measurements to the nearest millimetre and some surveyors therefore prefer to take dimensions to the nearest 5 millimetres, although all the dimensions in figures 1, 2 and 3 are taken to the nearest millimetre. It will be appreciated that not all the walls are vertical and hence horizontal measurements can vary according to the height at which they are taken. Hence it is good practice to work at a uniform height, normally just above window sill level. Ground and first floor plans may appear identical, but the position of internal walls and partitions on both floors must be carefully checked to ensure that they do coincide.

It is important to check that all measurements and details have been recorded in one room before moving to the next. These include the length and width of room; position of fireplace and size of breast; position of window(s) and the width, height and type; position, width, height and type of door(s); diagonal measurements, height of room, thickness of external and internal walls, and details of cornices, coves, picture rails, chair rails, dados, skirtings, floor finish, vertical pipes and fittings. The latter details are probably better recorded on a separate sheet or schedule.

The thickness of a party wall can be determined by measuring externally between windows and deducting the internal wall dimensions. Stud partitions can normally be identified by tapping them. It is also necessary to establish the direction of floor joists which will be at right angles to the floorboarding, provided it is not covered with parquet or other finish.

With old buildings, the surveyor should be on the lookout for straight joints externally which may indicate different dates of construction, jambs of blocked windows, evidence of doorways converted into windows and vice versa and heightening of walls. He should also note irregularities in plaster that may indicate the position of concealed timbers, blocked openings or the original jambs of a fireplace which has been partially filled in. Exposed beams, changes in ceiling level and other overhead features are recorded by broken lines on the plan.[17]

When sketching elevations, the recording of the number of brick courses will provide a good indication of the heights of chimney stacks and other inaccessible features. Brickwork normally rises four courses to 300 mm, but this needs checking on site. Upper floor thicknesses can usually be determined by measuring the floor to floor heights at staircases and deducting the lower room height and floorboarding and plaster thicknesses to the upper floor, otherwise it will need deducing from external vertical dimensions between windows compared with the appropriate internal dimensions.

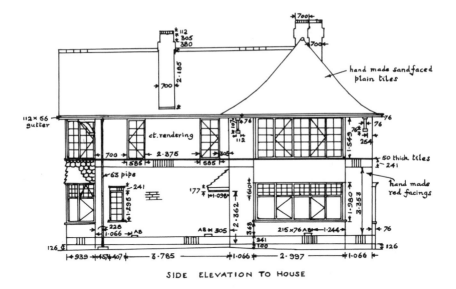

SIDE ELEVATION TO HOUSE

Figure 2. Measuring a building elevation

In addition to the preparation of floor plans, elevations and possibly sections, a site plan will be required. The site plan will show the position of the building(s) in relation to the site, any ancillary buildings, boundaries, driveways and paths, mature trees and manholes and lines of drains. Overall plot dimensions and diagonals at the corners will be required to plot the site boundaries as illustrated in figure 3. Diagonal measurements are also frequently used to position buildings on the site.

Plotting Floor Plans

Floor plans, elevations and sections are normally drawn to a scale of 1:100 and site plans could be 1:500. When plotting the floor plan of a house it is good practice to start at the left hand front corner, setting out each room until the skeleton of the building is complete. Once all dimensions tally, the insertion of further details, such as fireplaces, doors, windows, stairs and fittings, can begin. Leading dimensions of all rooms should be included.

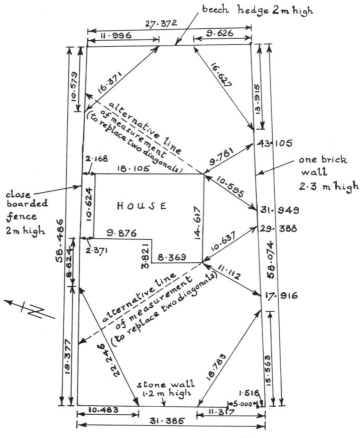

NOTE : All dimensions are metricated

Figure 3. Measuring for site plans

References

1. J. Bowyer. *Guide to Domestic Building Surveys.* Architectural Press (1979)
2. Royal Institution of Chartered Surveyors. *Structural Surveys of Residential Property: A Practice Note* (1981)
3. H. S. Staveley and P. V. Glover. *Surveying Buildings.* Butterworths (1983)
4. M. Hollis. *Surveying Buildings.* Surveyors Publications (1983)
5. PROTIMETER plc. Protimeter Technical Data Sheet. *How to Measure Moisture in Buildings* (March 1984)
6. T. M. Aldridge. *Boundaries, Walls and Fences.* 5th edition. Oyez Longman (1982)

7. N. Braine. Word processors for structural surveys. *Chartered Surveyor Weekly* (25 November 1982)
8. G. Biscoe. Responsibility for structural surveys. *Transactions of Royal Institution of Chartered Surveyors* (April 1953)
9. I. Melville. Surveyors limber-up for timber frame. *Chartered Surveyor Weekly* (12 January 1984)
10. Building Research Establishment. *Digest 251: Assessment of Damage in Low Rise Buildings.* HMSO (July 1981)
11. G. Edwards. Financial and legal aspects of subsidence. *Chartered Surveyor* (July 1979)
12. Building Research Establishment. *Report on Prefabricated Reinforced Concrete Houses.* HMSO (1984)
13. P. Eley. Use of redundant buildings: survey of information sources. *Architects' Journal* (21 March 1979)
14. M. Bussell. Use of redundant buildings: assessing the structure. *Architects' Journal* (21 March 1979)
15. Royal Institution of Chartered Surveyors. *Guidance Note on Structural Surveys of Commercial and Industrial Property* (1983)
16. Royal Institution of Chartered Surveyors and Incorporated Society of Valuers and Auctioneers. *Code of Measuring Practice* (1979)
17. R. W. McDowell. *Recording Old Houses: A Guide.* Council for British Archaeology (1980)

3 Building Defects

This chapter examines the principal building defects and their rectification, as they form an important part of any building survey. These are dealt with mainly in outline and readers requiring more detailed information are referred to *Building Maintenance*[1] by the same author.

The Building Research Establishment conducted a survey of building failure patterns and their implications[2] and the most common defects are shown in table 3.1. In a sample of 510 buildings, 58 per cent of the defects originated from faulty design, 35 per cent from faulty execution, 12 per cent from faulty materials, components or proprietary systems and 11 per cent from unexpected user requirements, but there was some overlap between these categories.

Table 3.1 Most common building defects

Building type	Defects (per cent)				
	condensation	rain penetration	cracking	floors	roofs
Council houses	59	13			
Council flats	38	33			
Private houses	18	33	20		
Private flats		64			
Factories		29	29		
Offices		28	19	28	23
Schools				32	39
Hospitals				35	26

Note: Shops, churches and universities showed no clear pattern.

Sub-structures

Foundations

General characteristics The foundations to an existing building are required to transmit the dead and live loads imposed on them over a sufficient area of soil, the area being influenced by the type of soil. Prior to the extensive use of concrete in foundations, brick footings were frequently used to spread the wall load over the ground, and the surveyor must be prepared for this with old buildings.

Unstable conditions in existing foundations could result from the breakdown of the foundation materials, such as by sulphate attack on concrete, but is much more likely to be caused by the movement of the surrounding ground. Occasionally a change in building use may result in heavier loads being carried by the foundations than was originally anticipated.

Main causes of defective foundations The main causes of defective foundations are now examined:[3]

(1) *Differential Loading*

Differential patterns of loading are likely to lead to differential settlement only where there are changes in ground support. The principal exceptions are bays, rear additions and internal partitions, which may be built on smaller and shallower foundations. There is often differential movement between these features and the main enclosing walls. In general, the deeper the foundation, the less the amount of consolidation. The method of bonding of bays and additions to the main structure will also have an effect on the extent of differential movement.

(2) *Undermining of Foundations*

Foundations may be undermined by the removal of ground support, as can result from mining subsidence or swallow holes. Mining subsidence is frequently accompanied by a surface wave which advances slightly ahead of the working face. A building on the crest of the wave will be in tension, changing to compression when it sags in the trough, resulting in probable fractures to the structure, sagging of arches, collapse of beams and fracture of pipe joints. Swallow holes are fissures in chalk and limestone which can swallow up significant quantities of soil. Other common problems arise from groundwater movement which undermines silts and peats, particularly near coasts and rivers, and clays which swell and shrink with moisture content and are affected by rainfall, leaking drains and water services and the demands of plant roots. The effect of clay soils is dealt with in more detail later in this section.

(3) *Building on Slopes*

Buildings erected on sloping ground are normally stable after a suitable period, although changes in groundwater could affect clays and gravels, whereby the building might settle and also slip down the slope. Changes in water content can affect made ground, and retaining walls may fail if subjected to excessive pressure.

(4) *Loss of Soil Support*

Loss of soil support may result from adjacent ground works, drain improvements and trenches excavated close to the building and only loosely backfilled, causing soil to slip towards them.

(5) *Chemical Attack*

Attack by sulphates or acid substances in natural ground is rarely sufficiently destructive to cause significant damage to shallow foundations of low rise buildings, since they are usually sited above groundwater level. Aggressive chemical compounds in fill material can, however, cause disintegration and expansion of ground-bearing slabs and foundation brickwork, particularly where burnt colliery shale has been used as under-floor filling.[4]

(6) *Consolidation of Poor or Made Ground*

In these circumstances progressive damage is likely to occur within the first ten years of the life of the building. It is advisable to obtain all available information about the history of the site from geological maps, old Ordnance Survey maps, aerial photographs and local residents. These sources will normally reveal for example whether the site has been in-filled or whether there were soft marshy conditions on the site originally.[4]

Shrinkage and heave due to clay soils   The Building Research Establishment[4] has identified the existence of clay soils as the major cause of foundation movement. Three distinct situations can arise:

(1) open ground away from major vegetation;
(2) buildings near existing trees; and
(3) buildings on sites recently cleared of trees.

In open ground, foundations less than about 1 m deep in clay soils are likely to be subject to seasonal movement, resulting in slight cracks which open and close seasonally. These may be unsightly but are easily masked.

Where existing trees near the building have been identified as the main cause of foundation movement, considerable care and skill is required in determining the likelihood of progressive movement and hence the form of remedial action. The Building Research Establishment[4] advises consideration of the following factors.

(1) Where trees have reached or are close to maturity, seasonal shrinkage and swelling movements can be expected, but larger movements will probably occur only in long periods of dry weather. Felling of these trees can result in worse damage because of swelling of the clay. Tree pruning may provide a method of reducing the harmful effects of the trees.

(2) Where the trees are far from mature, progressive foundation movement is likely to take place, causing increasingly severe damage. The only effective solution may be to prune or fell the trees. Relatively small changes in environmental conditions, such as leaking drains or water services, can lead to further root growth. The relative seriousness of the problem will depend on whether the structural damage extends down to the foundations.

Where buildings are erected on sites recently cleared of trees, the desiccated clay will absorb moisture and the ground will swell. This swelling may continue for many years and can cause progressive movement of foundations.

Inspection of foundations   The first source of information on foundations could reasonably be expected to be the plans deposited with the local authority under the Building Regulations or preceding Building Byelaws. Unfortunately, as described by Wilde,[5] many designers show the foundations on drawings of new buildings as provisional. For example, a note may be inserted to the effect that 'the foundations are to be agreed by the local authority on the site', or a break line may be drawn on the walls immediately above the foundations, or there may be no figured dimensions for works below ground.

Hence, as described by Pryke,[3] it is often necessary to dig inspection pits to check the depth, form and quality of foundations, possibly supported by specialist soil tests to establish the bearing capacity or to identify major faults such as swallow holes and mining subsidence. When digging pits it is important to disturb foundations as little as possible. Excavation is normally restricted to the points of greatest settlement, which often occur at the corners of buildings. The pits generally take the form of narrow trenches up to 600 mm wide, with digging stopping short of the bottom of the foundation and the last shovelfuls removed immediately prior to inspection. The inspection will be followed by careful and well-compacted backfilling, possibly using a fairly dry concrete around the foundation. The inspection may identify the need for costly underpinning.

Use of Hardcore

Hardcore is used as a make-up material to receive concrete ground floor slabs, raise levels and provide a firm base on which work can proceed. Ideally the material used should be granular, drain and consolidate readily, be chemically inert and not affected by water. In practice, few materials available at reasonable cost satisfy these requirements completely. Hence problems may arise through chemical attack by hardcore materials on concrete and brickwork mortar, settlement because of poor compaction and swelling or shrinkage resulting from changes in moisture content or chemical instability.[6]

Basements

Problems of damp penetration often arise with basements, aggravated by porous or flaking bricks or cracked concrete. On occasions the tanking of mastic asphalt or other waterproof membrane is applied to the inner face of the basement where it may be dislodged by the water pressure building up in the soil outside the basement. Some form of sandwich construction is likely to be most effective.

Remedial measures sometimes used with damp basement walls include the application of dry linings fixed on treated timber battens. This can prevent dampness and hygroscopic salts from damaging new decorations, but it will not provide a permanent cure. A polyethylene sheet vapour barrier should be fixed immediately behind the lining and it is advisable to ventilate the air spaces between the battens to the outside air. Bituminous dovetailed lathing (Newtonite) provides another useful alternative, fixed with rust-resistant nails and replacing the damp plaster.[1]

Retaining Walls

Retaining walls may be saturated for long periods and where constructed of bricks these should be resistant against frost and sulphates. Further basic precautions include the provision of adequate damp-proof courses and a suitable coping, the provision of land drains near the base of the wall on the retaining side and weep holes through the wall.

Walls

Defects in Wall Claddings

Wall claddings suffer particularly from the following defects:

(1) inability to support imposed loads, resulting in distortion or cracking;
(2) inability to keep out the weather;
(3) inability to insulate from the cold with resultant condensation; and
(4) deterioration of cladding materials.[1]

Cheetham[7] has described how the occurrence of defects in the fabric of a building can result from many unrelated design decisions: poor materials specification; inadequate assessment of loads; inadequate appreciation of conditions of use and inadequate assessment of exposure. Exposure is influenced by rainfall, direction of prevailing winds, the microclimate, atmospheric pollution, location, aspect and height of the building. The durability of building materials is also influenced by frost action, crystallisation of salts, sunlight, biological agencies, abrasion and impact, and chemical action and metallic corrosion.

Stability of Walls

Many gable end walls of large buildings were grossly under-designed in terms of the current Code of Practice (BS 5628) and some have collapsed as a result of high wind pressure and lack of intermediate support. Most of them had wall lengths in the 15 to 28 m range and mean heights of around 6 m.[8] Effective alternatives are the use of diaphragm and fin walls.

Cracked brickwork may not always be unstable. For example, the capacity of a 215 m brick wall to carry vertical loads may not be reduced by more than 30 per cent by a stepped or slanting crack up to 25 mm wide, provided that it is not accompanied by considerable transverse movement. However, the resistance to side loading of a half-brick wall with sound joints but with a visible bulge could be impaired considerably. With cavity walls the effects of leaning or bulging, and of eccentricity of loading, are far more serious than with solid walls, and wall ties play an important part in securing stability. Unfortunately, zinc-coated steel wall ties complying with the original BS 1243 have been found to have inadequate resistance to corrosion, and it became necessary to revise the standard in 1978 to require a considerable increase in the thickness of zinc or the use of plastic coating in addition to zinc.[9]

The significance of a defect must be judged in relation to the whole building – loadings, transverse support, openings and piers, all are important. It is also necessary to keep a sense of proportion – a wall which is out of plumb not more than 25 mm or bulges not more than 12 mm in a normal storey height would not usually require repairing on structural grounds.

Cracks in brickwork require careful examination, noting their position, direction, dimensions and any other characteristics, including whether they are progressive. A distinction is made between diagonal cracks which follow horizontal and vertical brick joints and those which pass through vertical joints and the intervening bricks. Diagonal cracks are filled with weak mortar, while the vertical cracks entail replacing the cracked bricks. With horizontal cracks it is necessary to establish whether the brickwork above the crack has risen or the part below has fallen. The cause of the cracks can vary, as for instance in the case of parapet walls, where cracking may result from expansion due to frost, thermal movement, sulphate attack or movement of the adjoining roof slab.

BRE investigations into a selection of prefabricated reinforced concrete houses,[10] revealed that the majority of reinforced concrete components are gradually deteriorating. They are doing so because of carbonation of the concrete and, in some cases, the presence of high levels of chloride, leading to corrosion of the steel reinforcement and the consequent cracking of the concrete. The great majority of houses studied were found to be in structurally sound condition, but there was a wide range in the rate of deterioration both between and within types. Some cracking was found in all types and the nature of the process is such that deterioration will continue, albeit very slowly in some cases, and all houses will eventually be affected by cracking.

Efflorescence

Very small amounts of salts, usually sulphates, which may be present in bricks and alkalis from the cement used in mortar, are sufficient to produce efflorescence during the period when a building is drying out. This is likely to be unsightly rather than harmful and should eventually disappear. It is destructive only in exceptional cases where the soluble salts crystallise just below the brick surface which, if weak, may crumble. Efflorescence should be dry brushed away before rendering a wall.[11]

Sulphate Attack

Sulphate attack on brickwork is the result of the reaction of tricalcium aluminate, present in all ordinary cements, with sulphates in solution. This causes expansion of the brickwork joints followed in extreme cases by their disintegration. The amount of soluble sulphates is limited in special quality bricks by BS 3921. Parapets and free-standing walls are the most vulnerable parts of the building, but in regions exposed to driving rain all external brickwork is potentially at risk if the other contributory factors are present.[11]

Dampness

The main sources of dampness in buildings have been identified by Oxley and Gobert[12] as direct penetration through the structure, faulty rainwater disposal, faulty plumbing, rising damp and dampness in solid floors. In this section we are concerned primarily with damp penetration through the structure and rising damp.

Penetration of damp through brickwork may, in particular, result from porous bricks, defective pointing, hairline cracks in rendering or lack of adequate weathering to external projections. South and south-westerly elevations of solid walls of buildings in the United Kingdom are particularly susceptible.

The introduction of cavity walls in domestic buildings reduced the incidence of rainwater penetration. However, it does still occur through mortar droppings in the cavity on wall ties and at the base of the wall extending above damp-proof course level, ineffective damp-proofing around door and window openings, and inadequate maintenance of rainwater goods. In each case the location and extent of the dampness will give a good indication of the source of the problem.

Rising damp can be a major problem, particularly with older domestic properties. Water in the soil rises naturally through porous building materials by capillary attraction. If unchecked, the rising water will pass through the foundations and upwards within the walls to appear as damp patches on the internal surfaces of walls. In newer buildings this should be prevented by the provision of effective horizontal damp-proof courses inserted above ground level. Unfortunately, these damp-proof courses can sometimes be by-passed by placing soil or other materials against external walls to a height above the damp-proof course. The erection of porches, sun lounges and other additions, often by do-it-yourself enthusiasts with little constructional knowledge, can also result in the damp-proof course being bridged and damp penetrating the main building. External rendering extending below the damp-proof course can be another cause.

The detection of damp by sight, smell and moisture meter is usually easier than determining the cause and deciding the remedy. In the case of older buildings the cause can often be traced to the absence of a damp-proof course, or to an inadequate or faulty one. In these cases dampness often shows as a tide mark along the length of the internal surfaces of walls. A fan-shaped patch on a wall may indicate a faulty damp-proof course in a particular location.

In time, dampness causes deterioration of plaster and induces the development of rot in adjoining timbers. Hence the renewal or replacement of plaster, floors and skirtings proceeds concurrently with the corrective treatment of rising damp. A new damp-proof course can be provided by silicone injection, syphonage, electro-osmosis, cutting out a mortar joint and inserting a traditional damp-proof course or cutting out a course of bricks and replacing with engineering bricks.

As described earlier, most problems of dampness are dealt with by a site inspection using visible signs of dampness and moisture meter readings to identify the cause of the problem. This is usually sufficient but it is sometimes useful to have confirmation by laboratory tests, even if only a simple salt analysis, to confirm rising dampness. Confusion can arise, particularly near ground level, where more than one source of dampness can result in a similar visible pattern. For example, rising dampness, condensation, rainwater splashback and natural drying out after replastering all concentrate their effects near skirting level and particularly in corners.[13]

Vegetation on the external faces of walls can cause problems. For example, creeping plants can damage masonry, lichens are useful for dating old stone but emit acid which corrodes metal flashings and gutters, while moss although picturesque, encourages insects, reduces evaporation from porous masonry and may block gutters. All plants should be trimmed to below eaves level and kept clear of window and door frames.

Cavity Insulation

The filling of cavities to hollow walls with urea formaldehyde foam has caused problems particularly when used with outer skins of porous bricks in exposed situations, through breakdown and shrinkage of the insulant and emission of gases.

Troubles have also been reported to the DOE[14] with cavity walls in new buildings where designers have met increased thermal insulation requirements by totally filling the cavity with mineral or glass fibre slabs as the walls are built. Experience indicates that, unless care is taken, these insulation batts can increase the risk of dampness internally, particularly in conditions of severe exposure. Problems can arise where slabs have to be cut or where wall ties occur at non-standard centres. Mortar 'snots' extruding from joints of the second leaf to be built can also cause trouble. Difficulties can arise above and around window openings and where they come into contact with damp-proof courses and cavity trays over openings. There may also be increased risk of frost and sulphate attack on the brickwork, unless materials of appropriate resistance have been chosen for the outer leaf.

Frost Action

In the United Kingdom, frost failures are mainly confined to brickwork in conditions of severe exposure, such as free-standing walls, parapets and retaining walls and, occasionally, work below damp-proof course. Bricks in these positions should have good frost resistance and be built in strong mortar to prevent spalling of the face of the bricks and disintegration of the mortar.

External Renderings

About 30 per cent of special investigations undertaken by the Scottish Laboratory of BRE related to rendering failures, a threefold increase in the period 1976–81, compared with 10 per cent in England and Wales.[14] In Scotland cement rendering traditionally provides a low-cost finish for masonry walls, but failures frequently occur, primarily owing to poor design and poor workmanship. Inadequate preparation and wetness of the background lead to a weak bond, and continual wetting, drying and frost action cause detachment of large areas of rendering.

Timber Frame Housing

A BRE report[15] concludes that there is no evidence that there are more problems or greater risks of major defects resulting from design or workmanship than in traditionally constructed houses. However, it emphasises that particular attention must be paid to the installation of cavity barriers against fire spread and vapour barriers, and preservative treatment of structural members. It is believed that more than 20 per cent of dwellings built in 1982 were of timber frame construction. The principal shortcomings found in practice relate to poorly stored timber exposed to rain, punctured vapour barriers and inadequate fire stops. The softwoods commonly used in timber wall frames are all subject to risk of decay when kept in a moist condition. The surveyor will experience problems in establishing the type and position of the moisture and vapour barriers, and the adequacy of fire stopping in existing timber frame dwellings.

Condensation

Nature of Condensation

Condensation has tended to become a greater cause of dampness in post-war dwellings than rain penetration and rising ground moisture. Warm air can hold more water vapour than cold air and when moist air meets a cold surface it is cooled and releases some of its moisture as condensation. Air containing a large amount of water vapour has a higher vapour pressure than drier air and hence moisture from the wetter air disperses towards drier air. This has a special significance since:

(1) a concentration of moist air as in a kitchen or bathroom readily disperses throughout a dwelling; and
(2) moist air at high pressures inside buildings tries to escape by all available routes to the outside, not only by normal ventilation exits but also through the structure, when it may condense within it.[16]

Condensation takes two main forms:

(1) surface condensation arising when the inner surface of the structure is cooler than room air; and
(2) interstitial condensation where vapour pressure forces water vapour through slightly porous materials, which then condenses when it reaches colder conditions.

The term relative humidity (rh) expresses as a percentage the ratio between the actual vapour pressure of an air sample and the total vapour pressure that it could sustain at the same temperature (per cent rh at °C). Air is described as saturated when it contains as much water vapour as it can hold — it is then at 100 per cent rh. If moist air is cooled, a temperature will be reached at which it will become saturated and below which it can no longer hold all of its moisture. This temperature is the dew point.[16]

The occurrence, persistence, extent and level of condensation are influenced by many factors, of which the most important are probably:

(1) number of occupants of the property;
(2) type of dwelling and construction;
(3) heat levels maintained in the property;
(4) type of heating;
(5) length of time the property remains unheated;
(6) degree of insulation;
(7) amount of ventilation; and
(8) prevailing weather conditions.[17]

A BRE report[18] estimated that over 1.5 million dwellings in the United Kingdom are seriously affected by dampness caused by condensation, and a further 2 million have slight condensation problems.

Causes of Condensation

There are two main reasons for the increase in frequency and severity of condensation — (1) changes in living habits and (2) changes in building techniques. More wives are now employed, often resulting in dwellings being left unoccupied, unventilated and unheated for much of the day. Moisture-producing activities such as cooking and the washing of clothes tend to be concentrated into shorter periods of time. Furthermore, washing and drying of clothes are often carried out within the main dwelling area instead of in a separate washhouse or fairly isolated scullery. Flueless paraffin heaters are sometimes used for background heating and they emit 1¼ litres of water vapour for each litre of paraffin burnt. Additionally, occupants have become more sensitive to slight dampness in their dwellings and frequently endeavour to maintain a high standard of decoration, so that local deterioration assumes greater importance.

Structurally, probably the most significant change is the disappearance of many open fires and air vents which provided valuable ventilation routes. Modern windows reduce ventilation flows and this may be further accentuated by draught-proofing by occupants. Solid floors without an insulating floor finish or screed are slow to warm, and modern wall plasters and paints are less absorptive. Flat roofs and newer forms of wall construction also need to be carefully designed if they are not to lead to increased condensation.[1]

Thermal insulation laid on top of the ceiling in the roof space is one of the most cost-effective methods of conserving energy, but it also increases the risk of condensation in the roof. A BRE digest[19] highlights the importance of adequate roof ventilation, preferably by the provision of openings in the eaves on opposite sides of the roof. When modernising older buildings, it is common practice to instal suspended ceilings supporting an insulating quilt on the top floor. This creates a cold roof where the temperature can be reduced below dew point, resulting in the depositing of moisture. Here again adequate cross ventilation is needed, requiring a 10 mm wide continuous open strip along the eaves on opposite sides of pitched roofs and a total ventilation area of not less than 0.004 times the plan area of flat roofs.[20]

Hollis[21] has described how the use of factory components in industrialised building systems often contributes to condensation. The reinforced concrete panel system of building provides cold wall surfaces and low levels of air circulation.

Surface condensation can lead to unsightly and unpleasant blue, green and black mould growth on walls, ceilings, fabrics and furnishings, which produce many complaints from occupants. On paint it may show as pink or purple staining.[22] Condensation within the fabric is slower to show up but may be much more serious in the long term.

Diagnosis of Condensation

Rising damp can be distinguished from condensation by the pattern and positioning of staining, while moisture penetration through cavity brickwork across wall ties also shows pattern staining. Gutters and downpipes must be checked for cracks, defective joints, blockages and the resultant leakage and water penetration. Roofs should be checked for defects and here again the type and positioning of staining is often a useful guide. Less obvious causes of dampness are slight weeping at pipe joints and wastes, and pinhole leaks in pipes.

Condensation frequently occurs as occasional damp patches in cold weather, although a sudden change from cold to warm humid weather may also cause condensation. Apart from investigating damp conditions, attention should be directed to the heating arrangements, possible use of portable oil or gas-fired appliances, ventilation, arrangements for drying clothes, means of dispersal of moisture from the kitchen, form of construction of floors, walls and roof, and whether there is any uninsulated pipework. Measurement of temperatures and humidities will show whether conditions favourable to condensation exist at the time of measurement. Suitable charts and useful calculations are contained in *Condensation in Dwellings, Part 1.*[16] A sling or whirling hygrometer is useful for this purpose and consists of wet and dry bulb thermometers. Protimeters, as described in chapter 2, can help in indicating the amount of moisture held beneath the surface of any material. A surveyor also needs the capacity to assess the reliability of information supplied by occupants.

Remedial Measures

The principal remedial measures consist of improved ventilation, insulation or heating, or a combination of these. If the relative humidity is excessive, the amount of moisture must be reduced or temperatures raised. Alternatively, the moisture vapour should be removed at source, preferably by mechanical means.

Experiments by the BRE Scottish Laboratory on blocks of flats indicated that the most effective remedial measures for condensation resulted from the improvement of thermal insulation by cavity fill and the provision of a group heating system.[23] Further research suggests that dehumidifiers can be successful in solving the problem of condensation in houses which are damp and warm, but the main obstacles to occupier acceptance are noise of operation and high running costs.[24]

Fungicidal washes are used to clean down and sterilise affected areas to produce a surface free of active mould growth. Fungicidal paints can then be applied as the final decorative finish. However, observations by the BRE indicate that cleaning and redecorating with fungicidal products alone is unlikely to provide a lasting cure for mould and should be used only as an adjunct to more permanent remedies.[24]

Thermal Insulation

In the quest for increased energy conservation, the depth of insulant, such as fibre glass, in roof spaces has often been increased to 100 mm, and the cavities of a considerable number of dwellings have been filled with an insulating material. In alteration and refurbishment work, external walls may be of solid construction and dry linings have been applied to upgrade their thermal insulation qualities.

It is the filling of cavities with urea formaldehyde foam that has caused the greatest controversy as, apart from the possible damp penetration through bridging of the cavity at shrinkage points and other weak points in the construction, there is a potential risk of harmful gas entering the building if the top of the wall is not sealed or there are gaps in the inner leaf of the cavity wall. For these reasons, the installation of the foam is now tightly controlled by BS 5618 under a registered firms scheme. The safeguards include a pre-installation survey to assess likely penetration of the foam into the building, preventive measures to ensure a continuous barrier between the foam injected into the walls and the occupiable sections of the building, sealing of vents and other gaps in the outer leaf and the fitting of sleeves around air vents penetrating the wall. Particular care is required in the assessment of system-built structures before cavity foam insulation is carried out.

Finally, immediately after installation the installer must inspect the inside of the building. Any foam should be removed from the occupiable areas and the hole through which it passed must be effectively sealed. If the contractor is advised of fume emission he must take speedy remedial action.

There are alternative materials available for use in the filling of cavities, such as polystyrene beads and mineral fibre. These have, however, been less researched and, as described earlier in the chapter, may give rise to problems in certain situations.

Roofs

Pitched Roof Coverings

Often the first indication of trouble is a damp patch on a ceiling or in the top corner of a wall. Localised leakage may occur as a result of defective flashings, cement fillets which have shrunk or broken away from adjoining surfaces, choked or defective gutters or slipped or broken slates or tiles. Deflective flashings need redressing, raking out the brick joints and rewedging and repointing the flashings. Zinc flashings may perish and become pitted in industrial atmospheres and are best replaced with lead, copper or other suitable flashings. Defective cement fillets should be replaced with metal flashings.[1]

Choked gutters need cleaning and checking to ensure that they are satisfactory. Eaves gutters may need resetting to falls and rejointing, and defective lengths replacing. Coating the internal surfaces of parapet and valley gutters with bituminous composition may extend their useful lives. The adequacy of tilting fillets and cover at junctions of gutter coverings with adjoining slating or tiling should be checked. It is also necesssary to check that effective horizontal damp-proof courses are provided under copings to parapet walls.

Slipped or broken plain tiles are readily replaced as the tiles are usually nailed only on every fourth or fifth course, and adjoining tiles can be lifted sufficiently to permit a replacement tile to be hooked over a tiling batten. The replacement of slates is more difficult as it entails removing the defective slate by cutting off the nail heads with a slater's ripper and fixing the new slate at its tail with a copper clip or tack bent over the head of the slate in the course below.

Rain and snow may penetrate a pitched roof because the slates or tiles are laid to too flat a pitch without increasing their lap. The problem is aggravated on exposed sites and the use of a flatter bellcast at eaves to improve appearance creates a vulnerable condition at the point of greatest rainwater runoff. Plain tiles for instance should never be laid to a flatter pitch than 40°. In severe cases it is necessary to strip the roof covering and to replace it with one suited to the particular roof pitch, as for example to replace plain tiles with single lap interlocking tiles on a 30° pitch roof.

The life of slates or tiles is dependent on a number of factors, including the physical properties of constituent materials and method of manufacture, climatic conditions, degree of pollution and method of fixing. Poorer-quality slates may have a life of up to 70 years while some of the poorer machine-made clay tiles may be restricted to 40 years on account of their laminar structure which is susceptible to freezing conditions. Concrete tiles may have a longer life but their colour is often bleached over a comparatively short period. Galvanised nails are unlikely to last the life of the slate or tile and are a poor investment.

Where large areas of slates or tiles are defective it is generally more satisfactory to strip and renew rather than to carry out extensive patching. With older buildings, problems sometimes arise through the manufacturers ceasing to produce certain single lap tiles. In older houses, sarking felt is rarely provided under the tiles or slates and battens and so rain or snow penetrating the roof covering has direct access into the roof space. In extreme cases it is necessary to strip the tiles or slates and battens in order to nail a layer of felt to the upper side of the rafters.

Flat Roof Construction

Whether flat roofs are constructed of concrete or timber they need to be laid to adequate falls and to incorporate a vapour barrier. A vapour barrier is designed to prevent moist internal air reaching parts of the external wall or roof construction which are cold enough to cause moisture vapour to condense. It is placed on the warm side of the insulation.

The generally accepted minimum fall is 1 in 80, but the latest recommendation in BS 6229[25] is 1 in 40 to allow for any inaccuracies on the site and possible deflection of the roof structure. The reputation of flat roofs for poor performance is partly the result of inadequate on-site workmanship and detailing and partly poor design.[26]

There are three principal forms of flat roof construction:

(1) *Cold deck design*, in which the insulation is below the deck, with a cavity between the two which must be effectively ventilated with air from outside the building and which remains at a temperature close to the external air temperature.

(2) *Warm deck design*, where the thermal insulation is supported on the deck and firmly secured to it by means of bonding or mechanical fasteners, while the waterproof covering is bonded to the top surface of the insulation. It is unnecessary to provide a roof cavity as a functional part of the roof, and there is no need for ventilation, which is an advantage over cold deck roofs where the ventilation may not always be effective.

(3) *Inverted design* consisting of a warm deck in which the insulation is placed over the waterproof membrane, which is thereby protected from solar radiation and extremes of temperature, and also from any traffic across the roof. However, this results in the weatherproof membrane not being immediately accessible for inspection and repair. No separate vapour barrier is required since the membrane, placed above the structural deck, also fulfils this function.[26]

Hence vapour barriers need to be provided in most roofs to protect insulation against water vapour from inside the building. With timber flat roofs the best approach is to introduce a vapour check at ceiling level, often in the form of aluminium foil backing to the plasterboard supporting mineral wool or similar insulation.[27] The roof space above the insulation should be ventilated, providing at least 300 mm^2 free opening every 300 mm on two opposite walls. The number of openings should be doubled if the span exceeds 12 m.[28]

Inspection procedures for identifying and diagnosing defects in flat roofs have been formulated by the Property Services Agency,[29] and can be summarised as follows:

(1) The examination of *records* to determine how the roof was built and under what conditions, and when it leaked.

(2) *Internal inspection* to look for staining, dripping, sagging, bulging, mould and smells, located by taking measurements to nearby features.

(3) *External inspection*, noting particularly falls and upstand heights, and examining the work in four categories:

    (i)   periphery: copings, upstands, parapets, gutters, trims and flashings;

   (ii) junctions: inadequate upstands, faulty jamb damp-proof courses and cracks;

  (iii) over supports: tears, splits, ripples, cockles, rucks and grooves over lines of beams; and

  (iv) body of roof: isolated defects, over panel joints, punctures, felt laps pulled open or surface deterioration.

(4) *Testing, sampling and opening up*: careful treading can identify soft patches, and the pumping of water out of hidden cracks; flooding can be used to test watertightness; deteriorating decks should be examined where possible from below. Specialist surveys may be considered necessary.

Bitumen Felt Roof Coverings

The early felts were based wholly on organic fibres which were very strong when new but dimensionally unstable, and could rot when moisture eventually penetrated the bitumen coating. The introduction in the early 1950s of bitumen felt based on asbestos fibre provided a more stable alternative, but not complete freedom from the risk of rotting. The development of bitumen felt based on glass fibre virtually eliminated dimensional instability and the risk of rotting but does not provide very high strength, so that the felt needs careful handling.[30]

The bitumen outer coating of bitumen felt roofing, if exposed to the weather, is gradually attacked by solar radiation. This deterioration can be postponed if the uppermost layer of felt, which is normally laid in three layers, has a surfacing of mineral aggregate, preferably white, partially bedded in a coat of bitumen dressing compound. Unfortunately, the mineral aggregate or stone chippings may in time puncture the felt and also obscure the source of leaks.

Splitting accounted for one-half of all bitumen felt roof failures investigated by the Building Research Establishment, and resulted from the inability of bitumen felt to withstand more than a slight amount of stretching without splitting or tearing apart. This defect may be remedied by patching with a strip of felt reinforced with hessian bedded in bitumen. The next most common cause of failure resulted from differential movement at skirtings to parapets and at other peripheral weatherings. Sometimes blisters develop between the layers of felt as a result of insufficient pressure being applied when rolling a layer of felt into hot bitumen bonding compound or the entrapping of moisture between two layers of felt. They do not often lead to leakage and no remedial action is usually necessary.

Upstands and skirtings should be integral with the surface felt and be formed by turning up the second and top layers against abutments to a minimum height of 150 mm. The felt should be turned up over an angle fillet at the base of the upstand to prevent the felt cracking at the bend or becoming damaged owing to lack of support. The angle fillet should be securely fixed to the roof to prevent distortion. Skirtings and upstands should ideally be masked by a metal or semi-rigid asbestos/bitumen sheet flashing.

Small holes in bitumen felt roofing can be sealed with a patch of felt bedded in bitumen. More extensive repairs may entail the removal of an entire sheet of felt by heating and softening a lapped joint and bedding a new layer of felt. Where general deterioration of the felt has occurred without fracture, a top dressing of hot bitumen and stone chippings may suffice.

Asphalt Roof Coverings

Two layers of asphalt are always necessary on flat roofs, with a finished thickness of 20 mm and joints staggered with a minimum lap of 150 mm between layers. Weak spots may occur if asphalt is reduced in thickness to obtain an even finished surface over raised parts, such as welted joints in flashings. Where a roof is likely to take considerable traffic, the asphalt is best finished with asbestos cement or concrete tiles, preferably with solar reflective properties.

A survey of 130 mastic-covered flat roofs to Crown buildings showed a 28 per cent failure rate, resulting either from splitting and cracking of the asphalt because of movement of the substrate and the absence of an isolating membrane, or peripheral cracking owing to differential movement between a roof deck and a non-integral parapet wall to which an asphalt skirting was fixed without any provision for movement. Slight hollows in a roof result in ponding and this may cause crazing of the asphalt but is unlikely to lead to water penetration. Cracked and blistered areas should be heated, cut out and made good with new asphalt without delay.[30] The new asphalt should be carefully bonded to the old by stepping the edges of the existing asphalt.

Problems have arisen through the application of white paint to the asphalt to reduce absorption of solar heat. The shrinkage of the relatively tough paint film is sufficient to pull the asphalt with it and cause cracking of the asphalt, with consequent loss of watertightness. Dampness in ceilings below asphalt roofs may result from interstitial condensation rather than moisture penetration through the asphalt.

Metal Flat Roof Coverings

Lead is susceptible to two forms of corrosion — in slightly acid conditions and under alkaline conditions, as with lime or cement mortars. Acidic conditions may occur through the discharge of rainwater on to the lead from pitched roofs containing algae, moss or lichen, but the corrosion process is generally very slow. A protective coat of bitumen made up of one coat of hot bitumen or two thick coats of bituminous paint in vulnerable locations should provide adequate protection. In general the resistance of lead to atmospheric pollution is very high, following the formation of a protective film of basic lead carbonate or sulphate, and may have a life in excess of 100 years.

Copper, like lead, forms a very effective protective film. Copper is resistant to alkalis but rainwater with an acid content dripping from algae-covered roofs or cedarwood shingles has been known to cause perforation of copper roofing within 40 years, and protection with bitumen is advisable in these situations. The majority of repairs to copper roofs are concerned with redressing rolls, seams and welts and repointing flashings.

Zinc has an average life of 40 years which is much shorter than that of lead or copper. Like lead, zinc is liable to corrode in both alkaline and acidic conditions. It is advisable to give at least one coating of bitumen to zinc embedded in plasters and mortars. Cracks can be repaired with bitumen and a surface dressing will prolong the life of a zinc roof.

Floors

Suspended Timber Floors

The most serious defect that is likely to occur in a suspended timber ground floor is an outbreak of dry rot, which is described in more detail in the next section of this chapter. It would entail the removal and destruction of all infected timber and also adjoining timber, together with the treatment of infected masonry. It may be necessary to improve ventilation by inserting new airbricks in external walls below floor level, preferably 215 × 140 mm terracotta at 3 m centres. Airbricks are needed on all walls to avoid stagnant corners and where suspended floors adjoin solid ones, it is advisable to lay air ducts under the solid floor. Sleeper walls under suspended floors must be honeycombed to permit a free flow of air under the timber floor. Internal partition walls will also need openings in them below floor level.

Any sources of damp penetration must also be dealt with, such as by the replacement of defective damp-proof courses and the lowering of outside ground or pavings where they adjoin or extend above the damp-proof course. All new timbers must be impregnated and it is advisable to treat existing timbers that are being retained. Where a defective suspended timber floor is constructed below ground level, the best remedy is to replace it with a solid floor incorporating a suitable waterproof membrane. In old buildings there may be no concrete oversite or damp-proof courses and full remedial work could be very expensive.

A suspended timber floor should not move perceptibly when walked upon nor should furniture or ornaments vibrate when placed on it. These deficiencies can result from inadequately sized joists or insufficient support to or fixing of the joists.

Where floorboards have curled across the grain, resulting in wide open joints (1.5 mm or more gap), the fault may be due to various causes such as use of unsuitable, wet or insufficiently seasoned timber, boards that are insufficiently cramped on laying, and insufficient nailing.

A large gap between the bottom edge of a skirting and floorboards generally stems from excessive shrinkage, both in the floor and the skirting. The squeaking of floorboards causes annoyance to occupiers and is usually due to loose fixing or the accidental contact of nails or screws with other metal components. Loose boards should be fixed more securely by nailing or screwing. In industrial buildings, badly worn floorboards may be lined with good quality plywood.[1]

Joists to upper floors are rarely attacked by dry rot but outbreaks can occur in damp cupboards or other enclosed spaces or through leaks in rainwater, waste or service pipes. More common faults in upper floors are sagging and springiness resulting from over-loading, inadequate size of joists or lack of strutting. Herringbone strutting should be provided to all floors with spans exceeding 3 m to stiffen the joists. A slightly springy upper floor can be strengthened by taking up floorboards and inserting solid strutting between joists, with each row of strutting wedged against the wall at each end. Another approach is to bolt new joists alongside existing ones, keeping the new joists shallower and with packing at their ends to avoid disturbing the ceiling. An old floor can be strengthened by screwing chipboard slabs over the existing floorboards.[1]

Solid Floors

There is always some risk of rising damp with concrete floor slabs supported on the ground. It is therefore customary with new floors to insert an effective damp-proof membrane to prevent possible damage to the floor finish. Different floor finishes offer varying degrees of resistance to dampness. For example, pitchmastic and mastic asphalt flooring both provide effective damp-proof membranes in themselves, while concrete, terrazzo and clay tiles transmit rising damp without dimensional, material or adhesion failure. Thermoplastic and PVC (vinyl) asbestos tiles may suffer dimensional and adhesion failure under severe conditions, while magnesium oxychloride, PVA emulsion/cement, rubber, flexible PVC flooring, linoleum, cork carpet and tiles, wood blocks in cold adhesives, wood strip and board flooring, and chipboard are all susceptible to damage in damp conditions.[31]

Concrete beds may be adversely affected by soluble salts in the ground or hardcore below. In severe cases it may be necessary to replace the concrete slab and the fill beneath, with a sheet of polyethylene between them.

Solid Floor Coverings

A *granolithic* finish is best laid monolithically with the concrete base. When repairing small areas of granolithic concrete, the defective material should be cut out in a rectangle with clean, straight edges, the exposed concrete covered with a cement slurry and the recess filled with new granolithic. In extreme cases it may be necessary to take up all the old granolithic and replace with new, adopting the generally recognised procedure.

Concrete paving may crack or wear unevenly. Cracks may be undercut on each side with a cold chisel to form a key for the cement mortar which is worked into the crack, after cleaning and wetting. Rough patches should be cut out and the exposed surface brushed and wetted, and covered with fine concrete to a minimum depth of 20 mm floated to a smooth finish.

Clay tiles may fail owing to arching or ridging as the tiles separate cleanly from the bedding, when the newly laid screed shrinks but the tiles remain constant. Where the tiles are firmly bonded by their bedding to the screed surface, considerable stresses may develop which are eventually relieved by areas or rows of tiles lifting. This normally occurs during the first year after laying and thin tiles rise more readily than thick ones. This defect can be avoided by introducing a separating layer of polyethylene or building paper over the base. If an old tiled floor becomes uneven or tiles are loose and cracked, the defective tiles should be removed, the surface of the base hacked to form a key, brushed and wetted, and covered with bedding mortar to receive the new tiles.

Thermoplastic and PVC tiles need polishing, preferably with emulsion polishes, to retain their initial appearance. Thermoplastic tiles are less resistant to oils and grease than PVC flooring.

Mastic asphalt and pitchmastic provide dustless, jointless and impervious floors but are liable to soften if in prolonged contact with fats, greases and oils.

Timber finishes take various forms. Boarding may be nailed to timber fillets which are either embedded in or resting on the upper surface of the concrete slab or screed. The fillets must be pressure impregnated with a suitable preservative and the upper surface of the concrete or screed effectively waterproofed. The underside of the boards should be treated as an additional precaution.

Hardwood strip flooring provides an attractive finish and requires sealing and polishing. Another alternative is wood blocks fixed with a suitable adhesive to a screed. The most common defects are unevenness, resulting from unequal wear which is normally cured by planing or sanding; loose blocks, often caused by shrinkage or expansion with subsequent loss of key, and cured by resetting the blocks in adhesive; and dry rot, caused by damp penetration and requiring removal of the infected blocks, treatment of adjoining flooring and renewal with suitably treated blocks on a substantial bed of bitumen or other appropriate material. It is advisable to leave an expansion joint between the edges of the block flooring and adjoining walls, often by means of a cork strip under the skirtings.

Staircases

The main defects found in timber staircases are worn nosings, creaking treads, cracked balusters and handrails, and loose newel posts. Cracked treads often result from the lack of angle blocks between treads and risers and the insertion of two or three angle blocks to each step should provide a cure. Cracked balusters and handrails are normally repaired by splicing and joining by screws or wood dowels. Loose newel posts can generally be stiffened by fixing angle brackets at their feet.

Timber Decay

The strength and usefulness of timber can be affected by a wide variety of defects, some of which occur during natural growth, others during seasoning or manufacture, while others result from attack by fungi or insects. The principal defects are listed and further information can be obtained from BS 565[32] and *Building Maintenance.*[1]

The principal defects arising from natural causes are knots, shakes, bark pockets, deadwood and resin pockets, while those due mainly to seasoning comprise checks, ribbing, splits and warps. Defects arising from manufacture are principally chipped grain, imperfect manufacture, torn grain and waney edges. However, the major problems arise from fungal and insect attack and each will now be considered.

Decay fungi are all limited in their activities to wood in which the moisture content exceeds 20 per cent. The moisture content of timber in most buildings varies between 10 per cent in well-heated and ventilated rooms to levels approaching 18 to 20 per cent in areas such as roof voids and suspended wooden ground floors, particularly in winter. Dampness may result from roof leakages, rising damp, condensation, penetrating damp or plumbing leaks.[12]

The most serious form of fungus is *Serpula lacrymans* (formerly known as *Merulius lacrymans*) which causes *dry rot*. Its special characteristic is to transport water through conducting strands and so, to a limited extent, moisten wood which would otherwise have been too dry to support fungal decay. It also has the ability to pass through lime mortars and plasters in its search for food. Timber that has been attacked shrinks and splits into distinctive cubical pieces by deep cross cracking. The dry rot fungus has a very strong and distinctive mushroom-like smell and forms large fleshy fruit bodies that are yellow to red/ brown in colour with white margins.

To eradicate an outbreak of dry rot, all affected timber and timber up to 300 to 400 mm beyond must be cut away, carefully handled and burnt on site. Surrounding masonry must be sterilised, preferably with a suitable fungicide, such as sodium orthophenylphenate or sodium pentachlorophenate. Ends of sound timber and all replacement timber must be treated with a suitable preservative. The cause of the outbreak must be established and rectified, usually by preventing damp penetration and improving ventilation. The most vulnerable locations are cellars, inadequately ventilated floors, ends of timbers built into walls, backs of joinery fixed to walls and beneath sanitary appliances. The design of timber floors to prevent dry rot is covered in *BRE Digest 18.*[33]

Wet rot was originally defined as chemical decomposition often arising from alternate wet and dry conditions, as occurs with timber fencing posts at ground level, probably accelerated by fungus attack. More recently the term has been applied to fungus attacks other than dry rot, for instance *Coniophora puteana* (formerly *Coniophora cerebella*) or cellar fungus, and the white pore fungus, *Poria vaillantii*. Other common fungi associated with wet rot are *Phellinus megaloporous, Lentinus lepidous, Poria xanthus, Trametes serialis* and *Paxillus panuoides.*[17]

Dealing with wet rot is a relatively easy and straightforward matter. Removal of the source of dampness and subsequent drying out will cure wet rot. Timbers which have become structurally unsound require to be replaced or strengthened.

Insect Attack

The life cycle of an insect is in four stages: the egg, the larva, the pupa and the adult. The egg is laid on the surface of the timber in a crack or crevice and the larva when hatched bores into the wood which provides its food. The larva makes a chamber near the surface and then changes to a pupa. Finally it develops into the adult insect and bores its way out through the surface. The principal wood-boring insects are now described.

Common furniture beetle or woodworm is about 2.5 to 5 mm long and dark brown in colour. Eggs are laid during July to August and larvae subsequently bore into the timber for a year or two, before the beetles emerge, mainly in June and July, each leaving a circular exit hole about 1 to 2 mm diameter; bore dust is in the form of ellipsoidal pellets. These insects prefer seasoned hardwoods and softwoods, both sound and decayed; also plywood made from birch or alder and bonded with animal glues is particularly susceptible.

Death-watch beetle is about 6 to 9 mm long, dark brown in colour and mainly attacks old hardwoods, especially in large structural timbers. The exit holes are circular of approximately 3 mm diameter and the bore dust is in the form of coarse, bun-shaped pellets.

Lyctus powder-post beetle is about 5 mm long and reddish-brown in colour. They favour sapwood in seasoned and partly seasoned sound hardwoods. They leave circular exit holes 1 to 3 mm diameter and bore dust is a fine talcum-like powder.

House longhorn beetle is about 15 mm long and is generally brown or black in colour. They are mainly encountered in parts of Surrey and prefer sapwood of softwood, especially in roof spaces. Exit holes are oval-shaped, approximately 10 x 5 mm in size, and large compact cylindrical pellets and powder are exuded.

Ambrosia (pinhole borer) beetle attacks fresh unseasoned hardwoods and softwoods, both heartwood and sapwood. Exit holes are circular and vary in diameter from 1 to 3 mm; bore dust is absent but the tunnels are darkly stained.

Wood-boring weevil is red-brown to brown-black in colour and attacks hardwoods and softwoods which are decayed or in damp situations. Exit holes are oval or slit-shaped with a ragged edge of varying size, 0.5 to 1.5 mm wide, and bore dust is in the form of ellipsoidal pellets smaller than those produced by full-grown furniture beetle larvae.

There are many proprietary preparations available for treating infested timber, most of which contain chemicals such as chlorinated naphthalenes, metallic naphthenates and pentachlorophenol. Insecticides should be brushed or sprayed over the surface during spring and early summer and injected into exit holes. The treatment should be repeated at least once each year during the summer months when beetles are active, until there is no sign of continued activity.

Joinery

Timbers for mass-production joinery are chosen mainly because they are inexpensive and easily machined, not because they are particularly resistant to decay. The Building Research Establishment[34] has identified two critical factors causing rapid decay — wrong choice of timber and water penetration at joints in rails, sills and other members. For example, Hemlock which is widely used for cheap windows is resistant to preservative treatment and is not even fully penetrated by the double vacuum process. Redwood which is used extensively in low-cost joinery includes a significant proportion of less durable sapwood but, fortuituously, it is readily penetrated by preservative.

Decay is not confined to the main members in a door or window. Dowels, glazing beads and plywood infill panels also give trouble. Glazing beads, not primed before fixing, and unprimed rebates encourage rot. Joints between members are often inadequately sealed.[35]

Windows

In recent years there has been a substantial increase in the number of instances of decay in wood windows in comparatively new houses. It occurs most frequently in ground floor windows and in the lowest part of the members concerned, such as the lower rail of an opening light, the bottoms of side frame members and mullions, and the sill, often at or near a joint. Wet rot is usually accompanied by discoloration and softening of the wood in its early stages and cross cracking later.[36]

Some window design details actually promote the entry and retention of water, such as with centre pivoted and top hung opening lights, left open during rain, in that the junction between glass and glazing beads collects water which will soak through joints into the woodwork. Early indication of conditions conducive to decay are given by putty failures, the waterlogged condition of wood beneath defective paint or discoloration of paintwork near joints. Open or strained joints and failure of paint over back putties may give rise to moisture penetration. Extensive swelling and jamming of opening lights indicate that moisture has gained access. Decay may first show as depressions in the surface of the wood or there may be wrinkling, discoloration or loss of paint. In these instances, the underlying wood should be probed with a bluntly pointed tool such as a small screwdriver to assess the extent of the damage.[1]

Where there is water penetration but no decay, remedial measures should be undertaken during a dry period of the year — stripping paint and defective putty, forming a slope on horizontal wood surfaces and working preservative into the joints of the woodwork, prior to reputtying and repainting. Where the woodwork is in an advanced stage of decay, the whole or part of an affected window may have to be replaced. Provision may also need to be made for collecting excessive condensed water on the inside of a window in a small channel and discharging it through weep holes drilled through the bottom rail.

Doors

Doors are sometimes a high-cost maintenance feature. Bottom rails of external doors are often devoid of weather ledges, resulting in rainwater collecting on horizontal surfaces and penetrating into poorly filled joints to promote decay. Plywood infill panels made from non-durable timbers and glues are liable to failure, and openings formed in them for glazed areas may allow water penetration. Joints between members of framed doors are liable to shrink, permitting the entry of moisture and possible decay. Doors which shrink excessively should be taken off their hinges and a strip fixed to the hanging stile.[1]

For the less severely attacked timber, the Forest Products Research Laboratory has developed an *in situ* treatment, embracing the injection of an organic solvent-type preservative, using special plastics injector plugs.[37]

Finishes and Decorations

Plasterwork

Ceilings and stud partitions may have lath and plaster surfaces. If battens to walls or laths are decayed, the plaster surface will be springy when pressed and will require stripping off and renewing. Laths may be decayed, become detached from wall timbers, be too close together so that the plaster lacks an adequate key or be fixed against wide components, such as large beams, so that there is no space behind for a plaster key.

It is relatively easy to identify poor plaster because of the hollow sound emitted when it is tapped. Disturbing plaster by chasing or pulling off wallpaper may result in large sections leaving the wall face. The quality of the plaster may vary considerably over a single wall and so it is inadvisable to replaster entirely after limited soundings. Generally two-coat work is sufficient when replastering, but three coats may be needed over very uneven walls.

Crazing to plaster surfaces may be sealed by painting with emulsion or liquid stopping. It is desirable to repair defective areas with plaster that expands slightly on setting, such as gypsum plaster class B. Small cracks should be cut back to a firm edge, undercut slightly to 3 or 4 mm wide. The enlarged cracks are wetted to remove dirt and loose material and reduce suction, and then filled with a skim coat or two coats if the crack is the full depth of the plaster. Where the surface is powdery or pitted it should be rubbed down, wetted and skim coated. Where large areas of laths are removed, the best solution is to fix expanded metal with galvanised nails or to fix plasterboard and skim-coat the new surface. If the walls are uneven and the mortar is soft, the best approach could be to apply expanded metal and render and skim, as the render would help bind the surface. With defective plaster cornices, patching is often preferable on grounds of both cost and appearance.

Other defects may occur in plasterwork, such as loss of adhesion of the final coat, which then requires stripping and replastering. Where efflorescence occurs, with soluble salts appearing on the plaster face, the surface should be carefully dry brushed as often as required. Flaking and peeling of the final coat generally results from persistent moisture penetration through the background, in which case the defective plaster should be stripped and a barrier to dampness provided or a cement-based mix used. Popping or blowing is remedied by filling small holes with a thick slurry of plastic paint or quick setting hemihydrate plaster, and larger holes by normal patching techniques. With recurrent surface dampness, strip the plaster and provide an impervious barrier. Soft or chalky plaster will normally require wetting, the application of a bonding agent and a suitable final coat of adequate thickness.

External Renderings

Smooth floated finishes are subject to surface crazing and this is accentuated with mixes that are rich in cement or which use fine sands. The render may become defective owing to structural movement. Spalling and lack of adhesion may result from poor workmanship, frost action and/or sulphate attack. Too strong a mix is likely to produce shrinkage cracks and loss of adhesion to porous or friable walls, accentuated by frost action. Sulphate concentrations may occur in chimneys or where a cement and sand mortar has been used for bedding or repointing, where the sulphates tend to crystallise, expanding off the render.[38]

Loose rendering should be cut away to a firm edge, undercutting it a little. It is advisable to cut away from about 100 mm around any movement cracks, wash out the gap and fill any cracks in the backing masonry. If structural movement is continuing or the masonry surface is friable, then it is best covered with expanded metal fixed with non-ferrous nails. The number of coats applied will depend on the thickness required and the outer one should not exceed 10 to 15 mm, using a 1:1:6 cement, lime and sand mix. Prior to subsequent painting it is advisable to use an alkali-resistant primer and undercoat.[38]

Painting

The Painting Process

Good appearance and durability demand good quality materials and craftmanship. Present painting costs rarely allow sufficient time for the proper preparation of the substrates. Although most present-day paints brush easily and permit the application of thin films, too much thinning of paint still occurs on site, to ease brushing and speed up the work. This results in thin paint films, poor finish and low durability.

Bare wood exposed to the weather is best left unpainted during wet periods or in winter and painted (after sanding or planing) when the moisture content falls. It can be protected temporarily by a liberal brush application of a paintable water-repellent preservative. Similarly, primers and undercoats subject to dampness should not be painted until they are dry.

There is an increasing use of exterior stains as an alternative to paint. They may not be as durable but their renewal is easier and cheaper and should be carried out as soon as they no longer show water repellency. Stains do not eliminate the need for preservative treatment, hide knots or imperfections, or protect putty, so they require good quality timber and bead glazing.[39]

Paint Failures

The main causes of paint failure are:

(1) crazing: undercoat has not hardened sufficiently before finishing coat is applied;
(2) chalking: destruction of oil paint by chemical or physical changes;
(3) blistering and peeling: painting on base with high moisture content, inadequate preparation or subject to excessive heat;
(4) wrinkling: too thick a paint film;
(5) discoloration: absence of knotting or painting over bitumen;
(6) blooming: delayed or defective painting; and
(7) saponification: alkali attack on paint, converting oil into soap.[40]

Repainting Woodwork

Attention should be concentrated on the vulnerable weathered areas of sills and lower rails. However, it is neither necessary nor economic to strip paint which is adhering well, chalking only slightly and free from other defects. The surface should be washed with a detergent solution or proprietary cleaner and preferably rubbed with wet abrasive paper. Any small areas of loose or defective paint must be scraped and sanded down to primer, if sound, or to bare wood and brought forward with primer and undercoat as appropriate before the remaining work is commenced. On sound existing paint, one undercoat and one finishing coat will normally suffice.

If the existing paint is soft, very chalky or eroded, cracked, blistered or peeling, or shows any adhesion weakness, it should be completely removed. Removal is also advisable if the paint has been affected by mould growth or by bleeding through of stains or preservatives, if there are already excessive coats. Exposed unpainted wood soon attains a fibrous or soft surface which offers insufficient adhesion for paint and must be removed by thorough sanding, scraping or planing. Decayed timber should be cut out and replaced; both old and new wood should be treated with preservative, particularly the end grain. *In situ* pressure injection of preservative is possible where existing joinery is liable to decay.[39]

The author in a paint and maintenance practice survey undertaken for the Paintmakers' Association,[41] found that in 1983 painting cycles were being extended beyond the normally accepted frequencies, particularly in the public sector, mainly because of shortage of funds. Furthermore, delayed painting had caused substantial and expensive deterioration of building components in many cases.

Repainting Metalwork

Repainting metalwork should not be delayed beyond the appearance of the first traces of rust. This avoids the more costly work later of removing rust and paint. The old paint surfaces can be rubbed down and finished with probably two coats of paint. Any very small patches of rust can be removed and touched in with an inhibitive primer. Complete removal of the paint followed by suitable surface preparation is necessary if rust covers more than 0.5 per cent of the area.[1]

Services

Plumbing

Pipes need to be securely fixed, properly jointed and free from leaks. Pipes and fittings should be accessible for purposes of examination, replacement and operation, but they are often concealed in older buildings. All valves need to be located and checked. Leaking taps waste water and are a nuisance, particularly with baths where they cause stains. Taps leak when the washers become worn or when metal seatings become eroded.

Most overflow pipes discharge in conspicuous positions so that the defect is soon noticed and remedial action taken. On occasions, however, overflow pipes discharge internally into fittings or are connected by hosepipes externally to the nearest gully. A ball valve may fail to close for one of several reasons — perforated float, eroded seating, defective washer or grit or lime deposit. The splashing of ball valves can often be reduced by fitting a silencing tube or drown pipe to the valve or by the use of Skevington/BRE controlled flush valves for WCs.[42]

Extensive damage to plumbing systems can occur in severe winters unless adequate precautions are taken against frost. These precautions include running the pipes in safe places wherever possible, fixing the pipes to falls with emptying cocks at all low points, lagging all pipes in vulnerable positions and the provision of ample background heating.

Many failures of galvanised steel tanks have occurred where they have been used in conjunction with copper pipes carrying aggressive water, which dissolves minute particles of copper. Electrolytic action between the copper and zinc can cause a rapid attack on the zinc coating. Rusting of the unprotected steel results in a leaking tank.

The avoidance of plumbing noises is often a matter of good planning and design and the choice of quiet appliances. When water flowing in a pipe is suddenly stopped by the rapid closure of a valve or tap, the pressure causes a surge or wave which rebounds from the valve and passes back down the pipe. The loose washer plate or jumper on the valve oscillates, producing a knocking sound known as 'water hammer'. The resultant pressure in small rigid pipes may damage them.

Modern buildings often incorporate single stack plumbing, in which certain criteria need to be observed for successful operation. A bath waste must enter the stack above or at least 200 mm below the entry of a WC branch. Wash basin, bath and sink wastes should have 75 mm traps to avoid the possibility of becoming unsealed and WCs must have a 50 mm seal. The maximum slope of a 32 mm wash basin waste depends on the length of waste pipe and where it exceeds 1.68 m in length it should be vented or alternatively a waste pipe of larger diameter should be used or a resealing trap. The bend at the foot of a stack should be of large radius. To avoid trouble from detergent foam, sanitary appliances on the lowest floor should be separately connected to the drainage system.[1]

Heating and Hot Water Supply

Defects in hot water supply systems may stem from one or more of the following factors:

(1) Air locks when air becomes trapped and impedes the flow of water. Trapped air is released by draining and refilling or by blowing through the pipework.

(2) Insufficient hot water caused by too small a boiler or hot water cylinder, excessive length of primary flow and return pipes, poor quality fuel, air locks, insufficient lagging of pipes and tanks, or possibly a combination of these defects.

(3) Noises may emanate from the primary flow and return pipes or the boiler, resulting from expansion of the water by freezing, furring or corrosion, and possibly involving descaling or renewal of pipes.

(4) Poor flow can stem from air locks, insufficient head of water or air drawn into the system through a vent. The latter defect can be remedied by inserting a larger cold feed pipe or raising the storage cistern.[1]

Back boilers for heating water are prone to leaks at about 15 to 20 year intervals and need checking. Immersion heaters and their thermostats also require checking. The thermostat is checked by switching off everything, setting the thermostat low and then letting the water heat up. When the meter stops, the thermostat is turned up and a check made to determine whether it has restarted. Feeling the water provides a crude temperature measurement.[38]

Old central heating systems often include cast iron radiators, which, although relatively inefficient, are uneconomic to replace since it would take many years before the cost of removal and replacement could be recovered out of reductions in fuel costs. Where hot water and heating are combined in a single system, independent supply of hot water is required outside the heating season and control of priority is useful within it.[38]

Smoky chimneys need checking for flue blockage, air starvation, poor fireplace and flue design, downdraught and other relevant factors. Closed-up fireplaces should have the chimney stacks suitably capped and the flues vented.

Air Conditioning

Air conditioning is becoming more common as some of the main advantages are more generally recognised. The design of air conditioning equipment is also changing and later forms of packaged equipment can be located in corridors, above false ceilings, on the roof or even in the conditioned space itself. Furthermore, the equipment is being designed to permit longer periods between maintenance periods. A low-cost sealed unit of limited life is a better buy than a longer life slow-turning open-drive machine which can be serviced.

Probably the most common cause of air conditioning plant failure is blocked filters, resulting in reduced air flow which can be followed by freezing of refrigeration plant and reheater batteries, and breakdown of fan bearings, causing downflow of cold air with consequent reduced occupancy comfort. Filters are of two main types – the throwaway and the cleanable, with the former proving the most popular.[43]

Electrical Installations

Modern electrical installations generally incorporate ring circuits; a recognised provision being one circuit for each 100 m^2 of floor area, and each ring can supply an unlimited number of 13 amp socket outlets. The ring system is cheaper and more convenient than the earlier arrangement, whereby each socket outlet required an individual sub-circuit with its own cable run from a separate fuseway. Spur connections may be taken from a ring circuit to serve outlying socket outlets with only two socket outlets or one fixed appliance fed from each spur, and not more than one-half of all points fed by spurs.

Socket outlets should be provided on a generous scale to give ample facilities for the use of electrical appliances without the need for trailing flexes and multiple adaptors. A minimum desirable level of provision is kitchen: 4, living area: 3, dining area: 2, each bedroom: 2, hall and landing: 1, garage: 1 and store: 1. Where two or more outlets are provided in a room they should, wherever practicable, be positioned on different walls and preferably be switch controlled.

Installations should be inspected and tested in accordance with the IEE Regulations at least once every five years. If repairs or alterations are required to electric wiring built in and concealed within the carcass or structure of a building, especially in domestic properties, it is necesssary to break open the wall or floor surface to gain access.

Wiring normally lasts about 30 years and aged circuits, installed before the early nineteen fifties can be identified by their round pin plugs and rubber sheath wiring which is probably perished, and a shortage of socket outlets. It is advisable to open up sockets and switches, particularly on damp walls, to check the cable type and to inspect for rust. Lighting circuits in older properties may also need replacing. With modern dwellings, particular attention should be paid to the extent of provision of external lighting and the more sophisticated switching devices and lighting fittings.

Gas Installations

Gas installations need to be regularly serviced and cleaned, and this aspect needs checking. The products of combustion must be discharged into the open air through suitable flues of pipes or blocks, of adequate size and soundly jointed. The surveyor when inspecting a gas installation should check for gas leaks by smell, and that visible pipes are not corroded and that all gas taps operate effectively. It should not be assumed that an existing supply is suitable for central heating. Existing pipes are often of too small bore and new supply pipes will be required. A simple check of heater efficiency can be undertaken by the surveyor by reading the rating, usually shown on the maker's plate, running the appliance on full for a few minutes and taking meter readings before and after running. Metered consumption divided by the time taken should match the rating. However, the problem is not usually that the appliance is inefficient but that it is of a type that is inappropriate for the anticipated use.[38]

Lifts

Lifts should be regularly serviced under a comprehensive maintenance agreement, a copy of which should be inspected. The surveyor should check on the adequacy of lift provision, having regard to the number of floors served and the probable number of passengers, and the degree of control. Lifts should be as vandal-proof as possible and any deficiencies in this respect noted.

Drainage

The joints on outlet pipes from appliances need checking to ensure that they do not leak and also that the traps are clear and the water seals sufficiently deep. Gratings should be taken off gullies and the traps cleaned of debris. Water can then be poured into them to establish the depths and adequacy of seals. Some old gullies have shallow seals and are therefore likely to emit foul odours. Rocking the gully gently will establish whether the joints are sound. Wet ground surrounding gullies may indicate leaks.[38]

Drains can cause trouble in a number of different ways. Loads from foundations of buildings or vehicles, tree roots or ground movement below drains can cause fracture of pipe joints or, in severe cases, fracture of the pipes themselves. Rigid cement mortar joints, used extensively in the past with clay pipes, are particularly vulnerable. Drains may also become choked through the deposition of silt or objects such as brushes and rags, particularly where the pipes are laid to flat gradients with restricted flows, and there may be no provision for access at changes of direction or gradient. Intercepting traps, now rarely installed, are another cause of blockage.[1]

Drains can be traced using a radio transmitter or surveyed by a television camera. The latter technique is expensive, costing at least £400 for the inspection of 700 m of drain in 1984.

Blocked drains are cleared by rodding, water jetting or winching from manholes and inspection chambers, and chemical cleaning is sometimes used for industrial drains. Defective pipes require replacing and leaking joints need cutting out and making good. Drains can be repaired by pressure grouting with either cement-based grout which is strong and rigid or a plastic gel which fills cracks and cavities and remains flexible. If leaking drains are suspected, they should be tested by water, air or smoke. Manholes and inspection chambers should be inspected periodically to check that they are in sound condition, particularly the benching and rendering, and that the drains are running freely. These chambers are liable to cause blockages, either because of their construction or because of disintegration due to age or chemical attack.[44]

Cesspools frequently leak and permit foul discharges into the surrounding ground and sometimes into nearby watercourses and even wells. The usual remedy is to waterproof the interior surfaces of the cesspool with asphalt, waterproofed cement mortar or a bitumen-based application such as synthaprufe. Septic tanks may on occasions require similar remedial treatment. Metal covers to manholes and other chambers may rust and require an application of bituminous paint and bedding in grease to prevent the escape of gases. Cast iron covers cracked by vehicles need replacing with heavier covers or possibly suitable steel covers.

Fire Precautions

Sound management, adequate fire protection equipment and effective fire prevention systems will reduce the likelihood of a serious fire, but they cannot eliminate it. Precautions to limit the spread of smoke normally include the provision of smoke-stop doors in corridors, at entrances to staircases and lobbies and in other suitable locations. Staircases should be ventilated by opening windows and/or skylights. Automatic dampers should be provided at strategic points on conveyors and in air conditioned ducts. In large single storey premises, roof ventilators assist smoke dispersal. Basements present a difficult ventilation problem and smoke outlets should be provided with fitted covers that can be removed or broken to permit the escape of smoke.[45]

The provision of adequate escape routes and fire alarms in the majority of buildings frequented by the public is required under various Acts of Parliament. Legislation in general provides for:

(1) provision and, where necessary, the enclosure of escape routes in suitable materials;
(2) aids to escape, comprising fire alarms, hand extinguishers and emergency lighting; and
(3) display of escape instructions and training of staff.[1]

References

1. I. H. Seeley. *Building Maintenance.* Macmillan (1976)
2. I. L. Freeman. Building Research Establishment. *Current Paper 30/75: Building Failure Patterns and their Implications.* HMSO (April 1975)
3. J. Pryke. Structural stability. *Architects' Journal* (23 June 1976)
4. Building Research Establishment. *Digest 251: Assessment of Damage in Low Rise Buildings.* HMSO (July 1981)
5. R. Wilde. Showing foundations on deposited plans. *Chartered Surveyor Weekly* (8 September 1983)
6. Building Research Establishment. *Digest 276: Hardcore.* HMSO (August 1983)
7. D. W. Cheetham. Defects in modern buildings. *Building* (2 November 1973)
8. Building Research Establishment. *Digest 281: Safety of Large Masonry Walls.* HMSO (January 1984)
9. K. E. Fletcher. Building Research Establishment. *Information Paper 1P 13/83: The Conformance of Some Common Building Products with British Standards.* HMSO (September 1983)
10. Building Research Establishment. *Information Paper 1P 16/83: The Structural Condition of Some Prefabricated Reinforced Concrete Houses of Boot, Cornish Unit, Unity, Wates and Woolaway Construction.* HMSO (October 1983)
11. Building Research Establishment. *Digest 165: Clay Brickwork: 2.* HMSO (May 1974)
12. T. A. Oxley and E. G. Gobert. *Dampness in Buildings.* Butterworths (1983)

13. C. T. Kyte. CIOB Technical Information Service No. 35. *Laboratory Analysis as an Aid to the Diagnosis of Rising Damp*. Chartered Institute of Building (1984)
14. Department of the Environment. Mineral and glass fibre slab cavity insulation. *Feedback Digest 40* (Autumn 1982)
15. Building Research Establishment. *Timber-framed Housing: A Technical Appraisal*. HMSO (1983)
16. Department of the Environment. *Condensation in Dwellings, Part 1: A Design Guide*. HMSO (1970)
17. A. Benster. Effective treatment of dampness in building. *Public Service and Local Government* (December 1979)
18. C. H. Sanders and J. P. Cornish. *BRE Report: Dampness: One week's Complaints in Five Local Authorities in England and Wales*. HMSO (1982)
19. Building Research Establishment. *Digest 270: Condensation in Insulated Domestic Roofs*. HMSO (February 1983)
20. Department of the Environment. Condensation over suspended ceiling. *Feedback Digest 40*. HMSO (Autumn 1982)
21. M. Hollis. Condensation: causes and cures. *Chartered Surveyor Weekly*. (25 August 1983)
22. Department of the Environment. *Condensation in Dwellings, Part 2: Remedial Measures*. HMSO (1971)
23. J. P. Cornish and C. H. Sanders. Curing condensation and mould growth. *BRE News 59* (Spring 1983)
24. Building Research Establishment. Dealing with condensation and mould growth. *BRE News 61* (Winter 1984)
25. British Standards Institution. *BS 6229: British Standard Code of Practice for Flat Roofs with Continuously Supported Coverings* (1982)
26. J.C. Beech and J. M. W. Dinwoodie. Preventing water penetration in flat and low pitched roofs. *Building Technology and Management* (November 1982)
27. Building Research Establishment. *Digest 180: Condensation in Roofs*. HMSO (1978)
28. Department of the Environment. *Advisory Leaflet 79: Vapour Barriers*. HMSO (1976)
29. Property Services Agency. *Flat Roofs: Technical Guide*. HMSO (1981)
30. Building Research Establishment. *Digest 144: Asphalt and Built-up Felt Roofings: Durability*. HMSO (1972)
31. Building Research Establishment. *Digest 54: Damp-proofing Solid Floors*. HMSO (1971)
32. British Standards Institution. *BS 565: Glossary of Terms Relating to Timber and Woodwork* (1972)
33. Building Research Establishment. *Digest 18: Design of Timber Floors to Prevent Dry Rot*. HMSO (1973)
34. Building Research Establishment. *Information Paper 1P 10/80. Avoiding Joinery Decay by Design*. HMSO (1980)
35. Department of the Environment. *Construction No. 26: Decay of External Joinery*. HMSO (1978)
36. Department of the Environment. Princes Risborough Laboratory. *Maintenance and Repair of Window Joinery*. HMSO (1972)
37. Department of the Environment. *Construction No. 24: In situ Preservative Treatment of Timber Window Frames*. HMSO (1977)
38. *Architects' Journal*. Repair and maintenance information sheets. (11 August 1976)

39. Building Research Establishment. *Digest 261: Painting Woodwork.* HMSO (May 1982)
40. I. H. Seeley. *Building Technology.* Macmillan (1980)
41. I. H. Seeley. *Blight on Britain's Buildings: A Survey of Paint and Maintenance Practice.* Paintmakers' Association (1984)
42. P. J. Davidson and C. J. D. Webster. Building Research Establishment. *Information Paper 1P 12/83: Water Economy with the Skevington/BRE Controlled Flush Valve for WCs.* HMSO (August 1983)
43. C. T. Gosling. Design considerations for easier maintenance of air conditioning equipment. *Building Maintenance* (October 1970)
44. R. Payne. *Drain maintenance: estate management.* Construction Press (1982)
45. British Insurance Association. The scale of the fire problem. *The Architect* (October 1973)

4 Structural Survey of Domestic Building

This chapter is devoted to a description of the work involved and the details recorded in carrying out a structural survey of a dwelling house, accompanied by supporting explanatory notes for clarification and amplification. The procedure and sequence of operations follows the approach described in chapter 2.

Background Information

Client: A. J. Smithson Esq. 'Courtlands', Cowslip Lane, Danby.

It is advisable to commence with the name and address of the client.

Property: 'Homelea', 28 Quebec Road, Broadhill, Suffolk.

Address of property for identification purposes.

Date of inspection: 24 October 1984.

The date of the inspection is important as there could be changes to the property subsequently.

Person(s) inspecting: George Halliway assisted by Roy Peters.

In a large organisation it is particularly helpful to know who carried out the survey, in the event of any queries arising later.

Weather conditions: Cold and dry, following a prolonged wet period.

Some defects are more likely to be clearly visible under certain weather conditions, such as damp penetration and leaking rainwater goods in wet weather.

Sources of information:
(a) Vendor supplied useful information about the history of the property.
(b) Local authority made original approved plans available and details of a subsequent submission under Building Regulations. The site was purchased from the local authority who also provided the rear service road.

It is useful to record any sources of information concerning the property and the type of information obtained.

History of property: The house was designed by a London architect and built for a director of a local factory in 1959. The property was purchased by the present owner in 1972, who had certain works carried out.

Age and alterations: The house is 25 years old and a second garage was constructed alongside the original garage in 1974. This work had local authority approval.

Structural repairs: The northern front corner of the house settled in the drought summer of 1976 and underpinning was carried out the following year. There is no evidence of further movement.

Flooding: Although Quebec Road was flooded when the River Stilwell overflowed in 1965, the house was not affected as it stands 5 ft (1.5 m) above road level. Since 1965 remedial work has been undertaken on the entire length of the river through the town, which should prevent any re-occurrence.

Rights of access: The owner has right of vehicular access on to the concrete service road at the rear of the site which is in the ownership of the local authority.

Tenure and rateable value: Freehold. GRV — £492.

The age of the property is significant, as are the number of changes of ownership. Design by a reputable architect normally ensures that the house is built to a good standard with materials of reasonable quality.

The type of construction, form of services and likely repair and replacement requirements vary with the age of the property. Confirmation is required that any major alterations have been approved by the local authority. It is advisable to note any major structural repairs that have been undertaken and their effectiveness. The existence of any guarantees, as with wood infestation treatment, should also be included.

Specific mention should be made of liability to flooding as this can be a serious matter for the owner in damage to the property and its contents.

Any special features such as restrictive covenants, disputes with neighbours and other unusual matters should be noted.

To complete the background information, it is usual to provide details of tenure and rateable value.

Detailed Survey

General Description of Buildings and Site

The house is detached and contains 5 bedrooms, lounge, dining room, kitchen, study,

A brief description of the buildings and site are provided

bathroom, 2 WCs, larder, boiler house and fuel store. The accommodation is very conveniently and compactly laid out with a minimum of circulation space. Sketch plans are attached showing the layout and sizes of rooms. The accommodation is as follows:

to give a broad picture which is very helpful to the client. Sketch floor plans showing the layout and sizes of rooms, as in figure 4, are also very useful. Dimensions are given in imperial

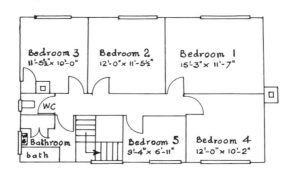

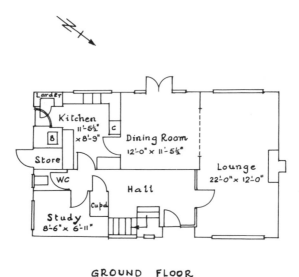

Figure 4. Sketch plans of surveyed dwelling

Ground floor

lounge – 22'-0" × 12'-0" (6.706 × 3.658)

dining room – 12'-0" × 11'-5½"
(3.658 × 3.492)

kitchen – 11'-5½" × 8'-9"
(3.492 × 2.667)

study – 8'-6" × 6'-11"
(2.590 × 2.134)

First floor

bedroom 1 – 15'-3" × 11'-7"
(4.648 × 3.530)

bedroom 2 – 12'-0" × 11'-5½"
(3.658 × 3.492)

bedroom 3 – 11'-5½" × 10'-0"
(3.492 × 3.048)

bedroom 4 – 12'-0" × 10'-2"
(3.658 × 3.098)

bedroom 5 – 9'-4" × 6'-11"
(2.845 × 2.109)

bathroom – 8'-6" × 6'-11"
(2.590 × 2.109)

The room heights are 8'-6" (2.490) on the ground floor and 8'-0" (2.438) on the first floor.

The house is built of brick cavity walls faced with hand-made multi-coloured facing bricks and the roof is of red sandfaced pantiles. Windows are of metal section set in timber frames.

Garages

Two single brick garages, built side by side, front on to the rear service road. The internal dimension of each garage are 20'-0" × 9'-3" (6.096 × 2.819).

Site particulars

The site has a frontage of 60'-0" (18.288) and a depth of 163'-0" (49.682). The site falls from back to front at an even gradient of about 1 in 28. The house has a north-east aspect and is set back 40 ft (12.19 m) from Quebec Road which was a busy road

measure at the request of the client and, although this approach might well be considered outmoded, it will be appreciated that this constitutes the normal practice. It is one of the disadvantages stemming from the lack of a universally accepted system of measurement with a mixture of imperial and metric measurements being used. Metric dimensions have been inserted in brackets to assist readers who are accustomed to working in metric measurements. They are entered in metres and millimetres, with the decimal point providing the dividing point between them, and thus eliminating the need to add metric symbols.

A brief description of the construction is helpful at this stage, although it will be described in more detail in later sections of the survey. It will be appreciated that the report will follow the same sequence as the survey.

A brief description of the garages is also given, to be followed by more detailed information later in the survey.

It is good practice to follow here with a description of the site, covering its dimensions and general characteristics. Information relating to external works and adjoining properties will be included later. The details con-

taking through traffic when the house was built, resulting in the provision of garages with rear access. Subsequently a by-pass road was constructed to the north of the town and Quebec Road has become a fairly quiet internal town road, although it does carry some traffic from an industrial estate some ½ mile (800 km) away.

cerning the construction of the by-pass is the type of information which is relevant to the particular site and is of special interest to the client.

The property is well located on a prime site in a pleasant residential area, and less than ½ mile (800 km) from the centre of the town. The town football ground is situated to the rear of the site, separated by a 6 ft (1.8 m) high concrete post and chain link fence on the south side of the service road.

The degree of accessibility to the main shops and other facilities should be given to assist the client in making a decision as to whether or not to purchase the property.

Roof void

It was possible to gain access to the roof space through a 24″ x 24″ (600 x 600 mm) access hatch. The roof timbers are of adequate size, in good condition and adequately framed together. They consist of 2″ x 4″ (50 x 100) rafters, supported by 1½″ x 3″ (38 x 75) collars and struts and 2″ x 7″ (50 x 175) ceiling joists, all at 24″ (600) centres. There was strong sarking felt under the tiling battens, and 1″ (25) of glass fibre insulation over the first floor ceiling, although it did not extend fully into the eaves. This should ideally be increased to 4″ (100) thick and to extend to within $\frac{3}{8}$″ (10) of the eaves to secure a significant reduction in heat loss and yet to allow adequate ventilation to prevent condensation.

There was no sign of damp penetration through the roof tiling, except for a small amount around the chimney stacks.

The roof timbers need careful examination for adequacy and condition. It could be that the timbers were of very poor quality, possibly poorly seasoned fast grown redwood containing large proportions of sapwood. Timbers of insufficient size or inadequately braced and supported could be distorted significantly and lead to water penetration. Roofs leaking over a period of time could result in decay of roof timbers through wet rot and there is always the possibility of infestation. With older properties the surveyor should be on the lookout for evidence of replacement timbers and, where this is so, to establish the cause. The extent of roof insulation should be noted.

There is a small area of boarding about 20 sq ft (1.86 m²) to provide a walkway but it is barely adequate.

Ancillary features such as walkway boarding deserve attention.

The cement mortar renderings to the chimney stacks, where they pass through the roof space, are crazing badly and need replacing.

Cold water storage tank

The galvanised steel cold water storage tank is unlagged and, worse still, badly pitted inside, probably as a result of electrolytic action caused by contact between dissimilar metals and the action of aggressive water. The tank requires early replacement, before it starts to leak. It has a capacity of 40 gallons (191 litres). The new tank should be lagged around the sides and be provided with an insulating cover. Polystyrene kits are available for this purpose. The tank has adequate timber supports and is provided with a satisfactory overflow. There is a valve on the downfeed pipe from the tank.

Room by Room Inspection

Bedroom 1

Plasterboard and skim coat ceiling — skim coat cracked over joints in plasterboard, which need filling in and the ceiling redecorating.

Walls — external brick inner skin to cavity wall and internal walls in concrete blocks, covered with gypsum plaster which appears to be in good condition, and wallpaper of good quality although badly faded. The joinery in $1'' \times 5''$ (25 x 125) skirtings, $3'' \times 1''$ (75 x 25) architraves and flush solid doors, $2'-6'' \times 6'-6'' \times 1\frac{1}{2}''$ (726 x

Chimney stacks in older properties will not contain flue liners and, in particular, those serving solid fuel boilers are likely to be attacked by carbonic and sulphuric acids, causing distortion of the stacks.

While in the roof space, the surveyor should examine all services to check their condition. Unlagged tanks and pipes constitute a potential risk and storage tanks should be examined for possible corrosion. Where pipes are lagged, the pipes should be temporarily exposed in parts to reveal the material used and its condition. Sometimes roof voids are festooned with cables and pipes and filled with surplus household goods, making a thorough inspection difficult. In this survey more information has been recorded than is often the case in practice, with a view to helping students to visualise the details and advice that will be needed in framing the report to the client, and also for future reference.

For ease of identification it is desirable to record the internal construction and condition on a room by room basis, even although this results in a considerable amount of repetition. This approach can be simplified by the use of a *pro-forma* or check list, normally prepared in schedule form, whereby brief notes can be inserted in the appropriate columns. The defects

2040 x 40) with aluminium alloy door furniture, are all in good condition, but the joinery needs repainting.

The flooring consists of 1″ (25 mm) tongued and grooved boarding supported on 2″ x 8″ (50 x 200) joists at 18″ (450) centres, with the joists prevented from buckling by the inclusion of herringbone strutting; it is of good quality and in sound condition.

There is a damp patch on the left hand reveal of the window opening, indicating that there is probably a defective vertical damp-proof course which requires investigation. Window boards are 1″ (25) softwood with rounded edges.

Bedroom 2

The finishes and joinery are similar to those in bedroom 1. The room has recently been redecorated and the walls are finished with emulsion paint. The ceiling is free from cracks. The door furniture is defective and requires repairing or replacing. The concrete block partition between bedrooms 2 and 3 is 2″ (50) out of alignment in its height of 8′-0″ (2.438); this is not considered serious as it is not carrying any significant roof load and probably originates from when the house was built.

Bedroom 3

The finishes and joinery are similar to those in bedroom 1, with the addition of a built-in cupboard between the projecting flue from the boiler and the end external wall. The condition of the plaster and decorations (emulsion paint) is reasonable, but dampness has penetrated down the outside of the flue through the roof space, resulting in a damp patch in the ceiling plaster and some decay (wet rot) in the studding above the cupboard door, where some poor quality untreated timber has been used. Replacement

can be grouped together under the various elements such as internal finishes, flooring and joinery in the report. It is advisable to record the nature and sizes of components wherever appropriate. In this instance it was found practicable to lift floorboards in order to determine the form of construction of the upper floor joists. This may not always be practicable.

The surveyor should always look for damp patches around window openings, generally owing to defective trays or damp-proof courses.

The form of the finishings and joinery are similar to those in the first bedroom inspected and so it is possible to reduce the amount of information recorded by referring back to the previous particulars.

A similar approach is adopted for bedroom 3. In this case some damp has penetrated into the bedroom and it is necessary to identify the cause, recommend action to prevent its re-occurrence and to make good the damage in the bedroom.

STRUCTURAL SURVEY OF DOMESTIC BUILDING

of the defective plaster and timber is required in addition to the rectification of the source of damp penetration which is covered later in the report.

Bedroom 4

The finishes and joinery are similar to those in the previous bedrooms. They are in excellent condition and the bedroom has recently been redecorated throughout (walls and ceiling in emulsion paint).

A brief entry is sufficient in this case because of the similarity of particulars and lack of defects.

Bedroom 5

This, the smallest bedroom, has been neglected because of its use as a boxroom. There are some cracks in the wall plaster and the decorations are in a poor state. The door and two lengths of skirting have been damaged. A small area of floorboarding has been attacked by furniture beetle, identifiable by the 1.5 mm exit holes and the gritty bore dust. The floorboarding has probably been attacked by beetle in old infested furniture stored in the room. The floorboarding requires thorough preservative treatment to prevent further insect activity.

The small bedroom has been neglected and thus warrants close examination. There is always danger in bringing old and possibly insect-infested furniture into houses, without first closely scrutinising the articles concerned. Having identified the type of insect, it is necessary to recommend the appropriate remedial treatment.

Bathroom

The plasterboard and plaster skim coat ceiling contains movement cracks resulting from changes of temperature and condensation. There is also some mould growth at one corner, requiring remedial treatment. The walls are tiled with ceramic patterned tiles to a height of 3 ft (900 mm) and plastered and matt oil painted above. The tiling is in good condition.

The window frame is showing signs of wet rot owing to the lack of a protective paint film.

The boarded floor is covered with linoleum which is badly worn and needs replacing, preferably with a different finish such as rubber or vinyl sheeting, or cork tiles. There is an airing cupboard in the bathroom

The conditions in a bathroom are often aggressive and the surveyor must be on the alert for the effects of excessive condensation on internal finishes and joinery. The client will probably be very concerned about the type and condition of the finishes and fittings as fashions and attitudes have changed considerably in the last twenty years.

Details of the airing cupboard and its components must be recorded, and the method of operation of the towel rail. Towel rails are often served from the hot water central heating

5'-3" x 2'-0" (1.600 x 600), containing three slatted shelves and a copper cylinder.

The sanitary appliances comprise a pampas ceramic bath, with mixer taps and shower, and pedestal wash basin to match. The bath is badly scratched and the shower is defective. There is also an electrically controlled towel rail which operates efficiently.

WC

A separate WC compartment contains a low level suite in white ceramic ware. The pan is chipped and the black plastic seat is of poor quality. The suite should preferably be replaced with a more modern fitting with a quieter valve. The plastered walls and ceiling are sound but need redecorating. The door lock is defective. The boarded floor is covered with linoleum which is in fair condition.

Landing

The landing is carpeted and in good decorative condition.

Lounge

The textured ceiling is crazed in parts and the decorations are in a poor condition. The patterned wallpaper is badly faded and marked in parts and some cracks in the plaster are visible through the wallpaper. The Affromosia wood block floor is in sound and solid condition apart from a few loose blocks near the doorway into the hall.

There is evidence of dampness around the south-west corner of the room, giving a fairly high reading on a moisture meter, and its location suggests mortar droppings on wall ties in the cavity.

The 1" x 6" (25 x 150) Affromosia chamfered skirting is in good condition. Attractive and soundly constructed oak panelled sliding doors separate the lounge from the dining room, and an oak glazed screen and single oak flush solid door separate the lounge from the hall. The screen is in good

system and will then be unheated in the summer months.

A separate WC is always an advantage, particularly in a 5-bedroom house. The condition of the finishes and type and condition of the appliance must be noted. Most prospective purchasers expect coloured suites nowadays.

Any points worthy of note about the first floor landing should be recorded.

The room by room inspection then continues throughout the ground floor, taking note of ceiling, wall and floor finishes, and joinery work. Electrical and central heating services and fittings are left to be taken later in their totality, although, as an alternative, they could be noted as each room is inspected to avoid the need for several tours of inspection. The damp penetration may require further investigation, such as the removal of facing bricks to locate the damp source and remedy it.

The glazed screen and sliding doors deserve particular mention as they are attractive and expensive features.

condition and well constructed with oak glazing beads fixed with brass cups and screws to a $4'' \times 3''$ (100×75) rebated and rounded frame and supporting tinted Cathedral glass.

An open fireplace is located centrally on the end wall, with a tiled hearth and surround. The tiled surround is set within a polished oak mantle shelf and jambs. A test of the fireplace showed a good air flow. The fireplace is sound apart from a cracked back firebrick. The flue is provided with concrete flue liners.

Dining room

The plastered ceiling and walls have recently been decorated and are in good condition. It has a wood block floor of similar construction to the lounge and is in sound condition apart from some scratched surfaces near the external doors. A pair of aluminium doors between sidelights lead to the paved terrace at the rear of the house. The doors are wind and weatherproof but slightly pitted. The door to the kitchen is a solid flush door faced with oak veneers, and surrounded by an oak architrave.

Kitchen

The plaster skim coated ceiling is very discoloured. The walls are fully tiled between the fittings. There is a stainless steel double drainer sink unit (hot and cold) with mixer taps, set in an L-shaped 'English Rose' range of working surfaces with cupboards and drawers beneath. There are full height wall cupboards on the remaining two walls; all in reasonable condition. The linoleum covering the floor is badly worn and needs replacing. The kitchen is rather cramped.

Leading off the kitchen is a larder and boiler room adjoining a passage leading to the partly glazed back door. The larder is $4'\text{-}0'' \times 2'\text{-}0''$ (1.220×610), plastered and emulsion painted, and contains three sets of shelves running continuously around three

All appropriate details of the fireplace should be recorded. The fireplace could prove to be an attractive feature to the prospective purchaser as they are now becoming popular again. It is important to check that it will function satisfactorily.

The same sequence is adopted for the dining room of ceiling, walls and floor, followed by joinery items. The condition of all components is particularly important, as this will determine the extent of replacement and/or remedial work and the expenditure involved. If the expenditure is very high it could result in the client deciding not to proceed with the purchase.

The particulars relating to the kitchen will include details of the cupboards, working surfaces and sink, and reference to their condition. It is considered advisable to make reference to the rather cramped condition of the kitchen which will prevent the later inclusion of any additional fittings or appliances.

Ancillary accommodation leading off the kitchen is best taken next. The particulars include the sizes of the compartments, their constituent materials and condition and the components con-

walls. The boiler house is $4'-0'' \times 3'-0''$ (1.220 × 915) and houses a gas boiler which replaced the original solid fuel boiler. The walls to the boiler room are of fair faced brickwork, and the floors to both the larder and the boiler room are of granolithic concrete, in fair condition. The doors are solid flush painted doors. The paintwork to these doors and the back door is in poor condition.

tained within them. Further details of the boiler are included in the services section of the report.

Study

The ceiling is plasterboarded, skim coated and lined, with a matt oil paint finish which is in good condition. The walls have been recently wallpapered. The floor is carpeted and the carpet is worn in parts and there has been some settlement of the concrete floor slab at the north-east corner, probably owing to inadequate compaction of the fill under the concrete slab.

There is a solid flush oak faced door leading to the hall, and a half-solid flush oak faced door giving access to a cupboard, both surrounded by softwood linings and oak architraves. The cupboard under the stairs is $3'-3'' \times 2'-9''$ (990 × 740) and contains a $10'' \times 1''$ (250 × 25) shelf. There are also fitted bookshelves $4'-3''$ (1.295) long, extending the full height of the room, with six shelves, made up of $9'' \times 1''$ (225 × 25) oak framing, soundly jointed together.

The same sequence is followed with the study and particulars of fittings included in the survey notes. Special attention must be paid to any visible defects such as the settled concrete and an attempt made to diagnose the cause of the defect.

WC

There is a separate WC off the hall but no wash basin. It is an efficient low level modern green ceramic suite. The plaster and paintwork are in good condition but the linoleum to the floor is badly worn. The door is solid flush and painted.

The type and condition of the WC needs recording and the general condition of the compartment. The absence of a ground floor wash basin in a house of this size is a distinct disadvantage and deserves mention.

Hall

The hall is reasonably spacious with adequate borrowed light from the porch, lounge and and windows over the staircase. The ceiling and walls are plastered and papered and in good condition. The floor is in Affromosia wood blocks which are soundly bedded but showing signs of significant wear. The front door is of oak, panelled and partly glazed, with adjoining sidelight in oak, both soundly constructed, leading to an external porch in brick facings with a matchboarded and painted ceiling. There is a cupboard under the staircase 3'-9" × 2'-9" (1.044 × 740).

The description of the hall should include its general characteristics, condition and any special features.

Staircase

The staircase is of dogleg layout with two quarter space landings, 3'-0" (900) wide, and with 15 steps to a conventional pitch. The staircase is soundly constructed with no noticeable deflection, of 1" (25) treads and ¾" (19) risers with sound nosings and with tongued and grooved joints; amply supported by adequate strings. The handrail is of 1" × 9" (25 × 225) mahogany supported by 1" × 1" (25 × 25) wrought iron balusters. The soffit of the lower part of the staircase has a sloping plasterboard and painted finish.

The staircase particulars contain details of the layout and construction, including the materials and sizes of the component parts. Attention should be paid to the relative stability of the staircase and the treatment of the soffit.

Services

Cold water supply

The main supply from the Water Authority's main is in ¾" (19) light gauge copper tubing with capillary soldered joints, which enters the dwelling through the boiler room where there is a screwdown stop valve and a drain-off valve. A ½" (12) connection is taken to the kitchen sink from the rising main at first floor level. The remaining connections are taken direct from the mild steel cold water storage tank in the roof space through ¾" (19) and ½" (12) pipes. There are stop valves to control all individual ball valves

All services are examined and their main components listed, with details of the materials, sizes and condition. It is advisable to separate the principal services — cold and hot water, heating, electrical, gas and telephone.

It is necessary to trace the position of the cold water services and to identify any connections which are taken off the

to storage and WC cisterns. They were all tested and found to be in satisfactory working order.

The taps to the sanitary appliances leak slightly and need rewashering, and it might be considered desirable to replace them with more modern fittings. There was no evidence of any leaking services, and the pipework and joints are in good condition.

Hot water supply

The heating source is a Potterton small-bore conventionally flued gas fired boiler, rated at 81 000 to 91 000 Btu's per hour, which is in sound working order. A 1″ (25) copper pipe primary circulatory system connects the boiler and the indirect cylinder located in the airing cupboard in the bathroom. The indirect copper cylinder has recently been replaced and is 1′-6″ (450) diameter and 4′-0″ (1.200) high of 40 gals (160 litres) capacity with a drain-off tap, 3 kW immersion heater and 2″ (50) thick insulating jacket. It is of adequate capacity for the system.

A 1″ (25)/¾″ (19) copper secondary flow and ¾″ (19) secondary return connect the cylinder to the appliances on both floors and it operates very efficiently as it is a compact layout. There is also a drain-off valve on the lowest point of the system.

The provision of the immersion heater enables the water to be heated in the summer months without operating the boiler.

The Leonard's thermostatic mixing valve to the shower over the bath is defective and requires replacement.

Central heating

The central heating system has its energy source in the gas fired boiler, supplemented by a Sund trand small bore pump to generate the desired pressure. A 10 gallon (55 litres) capacity feed and expansion cistern is connected to the system in the roof space.

rising main. The existence or otherwise of stop valves and drain-off valves need determining and testing. Any leaking pipe joints must be located and noted in the survey.

A detailed description of the hot water supply normally starts with the boiler, which may be gas, oil or solid fuel, with its rated capacity and condition. The primary pipework connecting the boiler and cylinder needs examining and describing, followed by the cylinder, giving all essential details as in the accompanying survey notes. The adequacy and effectiveness of the cylinder and the secondary pipework require careful consideration. On old systems the pipes may be in lead which is heavy and unsuitable for use with soft or acid water. It is also necessary to distinguish between direct and indirect systems as indirect systems are less efficient with their non-circulatory distribution pipes.

Central heating systems form an important component in modern dwellings and require careful investigation. The type of heating source must be stated and both oil and solid fuel boilers

A programme controller (Potterton mini-minder) is located in the kitchen. These fittings are all in sound working order apart from the pump which shows excessive wear, is very noisy and requires replacement.

The pipework is of light gauge copper with capillary soldered joints, two of which are sweating and need remaking. It consists of a ¾" (19) flow and return circuit above the ground floor ceiling with ½" (12) connections to the individual radiators which are of adequate capacity. The pipes are insulated with plastic foam but this has disintegrated. There are adequate valves on the central heating system of fullway gate pattern and these are in good working order. Metal sleeves are provided where the heating pipes pass through walls and ceilings. The pipes are inadequately supported and additional supports should be provided to ensure a maximum spacing of 4'-0" (1.200) on horizontal runs and 6'-0" (1.800) on vertical runs.

A suitable room thermostat (Satchwell) is located in the dining room on an internal wall. Ideally, each radiator should have thermostatic control valves to provide optimum control and conserve energy.

Radiators are of the panel pressed steel type complying with the appropriate British Standard and supplied by Potterton. The flow pipe enters each radiator through a wheel valve at the base of the radiator and the return pipe leaves at the opposite end through a lockshield valve. Each radiator is provided with an air release valve and a draw-off valve.

The following schedule gives details of the radiators.

need storage accommodation for their respective fuels. The general layout of the system and its adequacy and probable efficiency all need considering.

Hence all the components must be detailed and their condition described and performance monitored. Regard must be paid to the ages of the component parts, from the viewpoint of replacement. The surveyor should be on the lookout for defective fittings and/or pipework, particularly with older systems. For example, the design of hot water radiators has changed dramatically since the last war from the cast iron finned variety to the modern panel type which may consist of either single or double panels.

The earlier hot water central heating systems operated on the principle of gravity circulation, and they were almost invariably sluggish or failed to circulate. Those that worked were characterised by a wide difference between the flow and return temperatures, so that the radiators became progressively cooler along the line. This system was superseded in the nineteen fifties by pump-assisted systems using smaller diameter pipes and often referred to as small bore systems. In the late nineteen seventies, microbore systems were being introduced with $\frac{3}{8}"$ (10) diameter pipes. An alternative to panel radiators is skirting radiators which occupy less space.

Room	Nr	Double or single	Length	Height
Bedroom 1	1	double	66¼" (1.683)	24" (610)
Bedroom 2	1	double	59½" (1.511)	24" (610)
Bedroom 3	1	double	59½" (1.511)	24" (610)
Bedroom 4	1	double	59½" (1.511)	24" (610)
Bedroom 5	1	double	52¾" (1.340)	24" (610)
Lounge	2	double	52¾" (1.340)	24" (610)
Dining room	2	double	25¾" (654)	24" (610)
Study	1	double	52¾" (1.340)	24" (610)
Hall	1	double	25¾" (654)	30" (762)

The radiators are of adequate size for the rooms in which they are located, but there are none in the kitchen and a small one on the first floor landing would be an advantage. The two radiators in the dining room are rusting badly and leaking, and require replacement.

The central heating system as a whole performed satisfactorily under test and has been well designed. The radiators are relatively free from corrosion.

Gas supply

The gas service pipe enters the house at the front and runs to a gas meter located in the hall cupboard. A light gauge copper pipe, $1\frac{1}{8}"$ (28) bore, is taken from the gas meter, reducing to suitable diameters to serve the gas fired boiler and a gas point in the kitchen. The pipes are in light gauge copper and the exposed pipe in the kitchen is chromium plated and fitted with an effective safety control valve. The installation has been subject to annual servicing by the local Gas Board and is satisfactory.

Electricity supply

The electricity supply enters the house through the boiler room which contains the meter and a switchfuse control unit. The cables are PVC insulated and sheathed and are protected by galvanised trays where buried behind plaster, and are on ring circuits. The cables, although 25 years old, are still in good condition and are considered

Other forms of heating include warm air systems which generally lack effective control and can give rise to staining problems. Floor grilles are vulnerable and ceiling grilles are less efficient.

Underfloor heating was installed for a limited period and did not prove very popular, largely because of the lack of control and the problem with warm floors. A further development was the installation of storage heaters which occupy a considerable amount of space and require skilful use to obtain the best results.

The nature and extent of the gas installation needs recording even although it is very restricted. The safety aspect is particularly important and the surveyor should be on the alert for old corroded service pipes with poor joints and should always check on the position with regard to servicing.

The nature and scope of the electricity supply needs recording, including the control and meter position and the materials, adequacy and condition of the service. With older properties the cables are often rubber sheathed and may not be on a ring main

to have an effective life of at least 15 years. The system has been tested and approved by the local Electricity Board.

Lampholders are of white heat resisting bayonet type but are becoming increasingly brittle and distorted and should all be replaced. Light swtiches conform to the appropriate British Standard and are all in good condition; those in the garages are surface metalclad.

Socket outlets are of 13 A white flush switched type and have been augmented in recent years to give adequate provision, as listed below.

Bedroom 1	2 double
Bedroom 2	2 double
Bedroom 3	2 double
Bedroom 4	2 double
Bedroom 5	2 double
Bathroom	shaving point
Landing	1 single
Lounge	2 double, 1 single, TV socket and clock point
Dining room	2 double and TV socket
Kitchen	2 double, connection for cooker and clock point
Study	1 double
Hall	1 single
Garages	2 double in each

External wall mounted lighting fittings are fixed near the front and rear doors of the house. They are badly corroded and need replacing.

The lights in the lounge are of circular fluorescent type, and those in the kitchen and garages are fluorescent tube 5'-0" (1.500) long, 80 W, ceiling mounted, and the tube in the kitchen will soon need replacing.

The electric cooker is built-in Tricity four plate hob and double oven with white cooker hood, supplied from a separate circuit, and is in fair condition. There are elec-layout, and will almost invariably require rewiring. Modern uPVC sheathed cable, used extensively since 1958, is durable and normally satisfactory unless heavily overloaded. Power sockets, light switches, pendants and fixed electrical appliances should all be examined. Immersion heaters should generally be on a separate 15 A circuit with a double pole switch and have an indicator light. A check should be made for frayed cables to switches, ceiling roses, socket outlets, immersion heaters and other appliances. Pendants should preferably be earthed and power sockets be of the switched variety.

The surveyor should be on the lookout for amateur extensions undertaken by occupiers, which may not conform with the Institution of Electrical Engineers' Regulations and could be dangerous.

A schedule should be supplied of all power and lighting points and it is advisable for an electrician to test the system by a meggar test on cable insulation, polarity tests at the outlets and a test for earth continuity.

The distribution board will show the arrangement of circuits and in a typical dwelling will often consist of the following:

cooker – 30 amp
ground floor ring main circuit – 30 amp

tric 2 chime door bells to both front and rear doors.

A schedule of lighting points follows:

Bedroom 1	2
Bedroom 2	2
Bedroom 3	2
Bedroom 4	2
Bedroom 5	2
Bathroom	1 (pull cord operated)
WC (first floor)	1
Landing	1
Lounge	2 (circular fluorescent)
Dining room	2
Kitchen	1 (fluorescent tube)
Study	2
WC (ground floor)	1
Larder	1
Boiler room	1
Porch	1 (exterior ceiling mounted)
Garages	2 x 1 (fluorescent tube)
Externally	2 x 1 (exterior wall mounted)

Telephone
A GPO telephone is installed.

first floor ring main circuit
— 30 amp
immersion heater — 15 amp
ground floor lighting circuit
— 5 amp
first floor lighting circuit —
5 amp.

The cables in the roof space should be subject to close scrutiny. Ideally they should be attached to the sides of joists with clips at 9" to 12" (225 to 300) centres, with changes of direction made in junction boxes.

Circuits are normally tested at the mains position by means of a megohmeter connected to the circuit terminals, with the main switch at the consumer unit or fuseboard turned off, to check on insulation resistance and possible deterioration caused by moisture, atmospheric conditions or age. Each ring main and lighting circuit is tested. A separate ring main circuit should be provided for every 1000 sq ft (93 m²).

External Elevations

Roof
The roof to the house is covered with 'Greenwood' red hand-made sandfaced deep dish clay pantiles, laid with a single lap to a pitch of 30°. The tiles are in good condition, free from lamination and laid evenly, but there has been a considerable growth of lichen, giving a greenish hue, particularly on the front slope which faces north-east.

The survey should incorporate details of the roof covering, including the slope of the roof, and then follow with any defects which are observed from ladders or using binoculars at ground level. With older roofs, the tiles or slates may be laminating badly

There are two broken tiles and five that have slipped on the rear slope and these need replacing or rehanging as soon as possible, to prevent the ingress of water into the building. There are some spare tiles stored in the garage. Some of the cement and sand pointing between the half-round ridge tiles has become loose and it is necessary to remove the defective material and repoint with new.

Two chimney stacks serving the lounge fireplace and the gas boiler, penetrate the roof covering and the pointing to the flashings has broken away, permitting rainwater to pass down the sides of the chimney stack and enter the roof space and the first floor ceiling, so causing damage to plaster, decorations and joinery in bedroom 3. The brick joints need raking out, the lead stepped flashings reinserting in the brickwork, and repointing done to make them watertight.

The precast concrete capping to the chimney stack from the boiler room is cracked and loose, allowing damp to penetrate the brickwork. The capping requires replacement. The brickwork to the stacks needs some repointing.

or the nails corroding, resulting in the need to retile or reslate the roof, which will be an expensive operation. Particular attention needs to be paid to valleys which may become choked or defective. Broken or slipped tiles are much more serious with pantiles than plain tiles, as they have only a single lap. Hip tiles sometimes slip and hip irons may be missing. Chimney stacks are vulnerable features as they are very exposed. The weatherproofing items at the junctions of chimney stacks and roof coverings require particular attention. Flashings can become detached and cement mortar fillets break away, allowing damp penetration into the building. In the case of older properties, the brickwork may become cracked and loose, putting the chimney stacks in a dangerous condition. Chimney pots may also become loose as the cement mortar flaunching at the top of stacks cracks and breaks away.

Barge boards and eaves

The roof is gabled at both ends and finished with $1'' \times 9''$ (25 × 225) barge boards and soffit boarding to protect the tops of the gable walls. The barge boards appear to have been constructed of poor quality timber, which has shrunk, and the members are badly twisted and cracked. This has caused the premature breakdown of the paint film and the penetration of rainwater into the timber. Substantial sections of the barge boarding are rotting and it is advisable to replace the whole with good quality timber amply treated with preservative.

The survey should indicate the shape of the roofs and whether they are hipped or gabled. In the case of gabled ends the finish at the top of the gable walls shall be described, and also the condition of the construction, which could consist of tiled or slated verges, barge boards or stepped parapet walls. In all three cases, defects can arise which need describing, together with the remedial work involved.

There are also two lengths of rotting 1″ × 6″ (25 × 150) fascia board, primarily owing to lack of painting over the years, which require replacement. The 12″ × 1″ (300 × 25) soffit boarding is in fair condition, apart from a few localised sags where it is pulling away from the bearers, and appears to have inadequate support.

Rainwater goods

The eaves gutters are in 4″ (100) cast iron half-round section and are of adequate capacity. They are supported at 6′-0″ (1.800) centres with cast iron brackets. Ideally they should have been supported at 3′-0″ (900) centres. Apart from the insufficient support, the gutters are laid to inadequate and uneven falls and this has resulted in considerable silting and overflowing. There are a number of leaking joints, the gutters are rusting badly and there are two cracked lengths. The paintwork is almost non-existent.

The gutters need cleaning out, defective lengths replacing, joints recaulking, the number of gutter brackets doubling, extensive wire-brushing and completely repainting. The overflowing gutters have resulted in some areas of brickwork becoming permanently damp, although very little appears to have penetrated to the inside.

In like manner, the four 3″ (75) diameter rainwater downpipes have been sadly neglected, with two cracked lengths, several leaking joints and extensive rusting. Replacement of the cracked lengths, remaking the joints, and thorough cleaning and repainting is necessary, before damp penetrates the house and causes more serious damage.

Brickwork

The external walls are constructed in 11″ (255) cavity walling with the external skin in faced brickwork and the internal skin in insulating concrete blocks. The cavities have not been filled with an insulant. The facing

This should be followed in a logical sequence with the form of the eaves construction, its condition and any remedial work required. Cast iron gutters are prone to rusting unless kept regularly painted, and PVC gutters tend to become distorted under heavy snow loads, sometimes accompanied by the cracking of the gutter brackets. The adequacy of the gutters and the falls to which they are laid require specific comment. It is a building component which is frequently skimped, despite the fact that it performs such an important function. The importance of painting externally on planned cycles, the length of which is influenced by location, should be highlighted, as factors such as high level of atmospheric pollution or proximity to the coast necessitate more frequent painting.

Neglect of painting of woodwork and metalwork can result in the accelerated decay of the substrates, apart from producing drab and unattractive buildings. The importance of the regular painting of the exterior woodwork and metalwork of buildings cannot be over-emphasised.

The external cladding requires close examination and the construction and condition need carefully recording. Brickwork should be examined for fractures,

bricks are 'Newdigate' hand-made multi-stocks laid in Stretcher bond in gauged mortar with recessed joints to provide attractive elevations. There is a four course plinth at the base of walls projecting $1''$ (25).

There are some cracked joints to the brickwork adjacent to the lounge chimney on the north-west wall, but these cracks are very old ones, probably caused by the heat from the lounge fire, and there is no evidence of any further movement in recent years. It is recommended that the cracked mortar joints be raked out and repointed with a suitable gauged mortar of similar colour and strength to the existing. Where the projecting brick chimney breast tapers into the flue on the north-west wall, it is covered on the upper sloping surface by roofing tiles bedded on to the tapered brickwork. Some of the tiles are cracked and others have become dislodged. These need making good to provide the necessary weather protection.

Some of the bricks to the head of the projecting brick-on-end course over the central feature to the front wall of the house are loose and deteriorating owing to frost action and need replacing.

Rendering

There is a painted cement rendered area to the front wall of the house between the porch and the large staircase window, measuring $5'\text{-}0'' \times 8'\text{-}0''$ (1.524 × 2.438), and this is badly crazed. The rendering shrunk on setting, resulting in the formation of numerous hair cracks through which moisture can penetrate by capillary action. It is probable that the rendering mix had too high a cement content.

The rendering needs hacking off, the brick joints raked out to form a key and a rendering of roughcast in gauged mortar applied which should remain free from crazing. Colour can be obtained through the choice of aggregate or by an application of cement paint.

spalling bricks and other defects. It is also necessary to determine the cause of the defect and to recommend the appropriate remedial action. It is desirable to match the existing work as far as practicable. The surveyor should be on the lookout for porous bricks and sulphate attack. With timber cladding the surveyor should distinguish between ship-lap boarding and feather edge weatherboarding. The type of timber and its condition are of considerable importance and will have a significant bearing on future maintenance costs, which should be communicated to the client.

Any special features such as plinths, over-sailing courses and corbels should be described, together with a note of their advantages where appropriate. Areas of tile hanging can be an attractive feature but they need to be fixed to treated battens.

Smooth, dense cement renderings are liable to craze and so permit moisture penetration. Experience shows that roughcast is much less vulnerable to crazing and that it is good practice to use weaker mixes incorporating less cement, as they are more flexible and porous, and therefore permit rainwater absorbed by the rendering to dry out in warm weather. Dense renderings containing hair cracks enable moisture to penetrate the cracks and become trapped behind the rendering.

Damp penetration

Reference was made earlier in the survey to internal dampness in the south-west corner of the lounge and the possibility of this being due to mortar droppings on wall ties in the cavity. An examination of the external face of the wall failed to reveal any other cause, and it is recommended that small areas of brickwork be cut out to permit examination of the ties, removal of any mortar droppings, and, at the same time, to check on the type and condition of the wall ties. The Building Research Establishment has found instances of failures of wall ties often owing to substandard galvanising.

Damp-proof course

The damp-proof course is of bitumen sheet with hessian base and lead and is providing an effective barrier to rising damp. The cement mortar pointing is breaking away in several places and needs raking out and making good. The grassed area alongside the north-west wall finishes within about 1″ (25) of the damp-proof course and the ground should be lowered and the grass reinstated to give a gap of at least 6″ (150), and so prevent any possibility of the damp-proof course being bridged.

Underpinning

A close examination of the front north corner of the house was undertaken, where underpinning was carried out in 1976. It is understood that the concrete strip foundation, about 2′-6″ (750) deep at this point, was adversely affected by the roots of an ornamental tree planted close to the house in the shrinkable clay subsoil.

The foundation was suitably reinforced with concrete underneath the original foundation and the tree removed. There is no sign of any further movement and it is considered that the defect has been adequately rectified, and is no longer a matter of concern.

The dampness in the lounge requires further explanation as it will in all probability be a source of worry to the client. Coupled with the recent spate of failures of wall ties evidenced by the Building Research Establishment, it would certainly be good policy to recommend opening up the outer skin of the cavity wall in carefully selected positions opposite the internal damp patches to permit examination of the wall ties in these areas.

It is important to investigate the type and condition of the horizontal damp-proof course as it forms a vital barrier to rising damp. With older properties the damp-proof course may have failed or there may be no damp-proof course at all. A minimum distance of 6″ (150) must be maintained between the damp-proof course and adjoining external surfaces to provide adequate protection and to comply with the Building Regulations. Exposed brickwork in projecting courses, parapet walls and similar free-standing features, require special attention, particularly if the facing bricks have a relatively high level of porosity. On occasions the most unsuitable bricks are used in very vulnerable positions as, for example, facing bricks with an absorption rate of 24 per cent in parapet walls. The fact that costly underpinning work has been found necessary could well be a matter of concern to the client and he will need reassurance that there is no likely possibility of a reoccurrence. Any progressive movement of a structure requires monitoring over a period of time.

Windows

The front window to the lounge has a 3″ (75) deep artificial stone surround and the member at the head of the window is cracked in the centre and the steel reinforcement is badly corroded. This member requires replacement as, apart from the disfigurement, it is structurally unstable.

The windows are pivot hung in steel section in 3½″ x 3″ (88 x 75) hardwood frames. The painting of the exterior has been neglected over a long period, with the result that the metal windows are rusting badly and much of the timber surrounds are devoid of paint. If allowed to continue the thin steel windows are likely to distort and crack the glass and the timber surrounds will become increasingly susceptible to rot.

The windows to the kitchen and bedroom 2 already contain small rotting sections of timber which need cutting out and replacing. If the timber surrounds had not been in hardwood, the rot would have been far more extensive. All woodwork and metalwork needs cleaning down to bare wood and metal and then given a full preparatory and 3 coat painting treatment using good quality oil paint. A number of putties need replacing where they have become loose or detached. The external window sills are constructed of two courses of roofing tiles set in cement mortar and they are in good condition apart from two loose tiles.

Having completed the examination of the structural elements of the building, attention should next be directed to the external joinery. This is often a source of problems and justifies close examination. In newer properties, it is quite common to find that the windows are made up from fast grown redwood containing large quantities of sapwood, insufficiently seasoned and not treated with a preservative. In consequence, the timber shrinks and distorts, causing the thin paint film to crack. Rainwater penetrates the cracks, becomes trapped behind the paint film and decay of the timber follows in an astonishingly short period of time. The decaying timber often goes unnoticed by the owner or occupier who is easily misled by the apparent good condition of the paintwork.

External doors

The fuel store was converted into a garden store when the solid fuel boiler was replaced by a gas fired boiler in 1978. The external

In this instance both the rear and garden store doors are defective. It is advisable to

door is a softwood matchboard, ledged and braced, 2'-6" x 6'-6" (726 x 2040), with $\frac{5}{8}$" (15) thick tongued and grooved matchboarding, V jointed on both faces, nailed to 4" x 1" (94 x 22) ledges and braces, hung from a 4" x 3" (100 x 75) rebated softwood frame. The door is in poor condition with split boarding and the door has twisted and dropped at the opening edge. It is advisable to replace this door with an exterior quality flush door.

The back door is a semi-solid flush door 2'-6" x 6'-6" x 1¾" (726 x 2040 x 44), with the upper section glazed and beaded. Rainwater has penetrated between the beading and the bottom edge of the glass, causing rotting of the bottom part of the door, and the weatherboard is missing. The rot is well advanced and the door requires replacement, preferably with a panelled door, as this will be less vulnerable to wet rot, and a glazed section is necessary to provide natural lighting to the passage between the larder and the boiler room.

The concrete step, 8'-0" x 2'-6" x 9" (2.438 x 726 x 225), serving the back door and the door to the garden store is badly crazed and crumbling and appears to have been constructed of very weak concrete. This step requires replacement because of its dangerous condition. Front door was taken with hall.

describe their construction and dimensions, and the defects that have arisen with their consequences. Where it is proposed to replace defective components with a different type of unit, the reasons for this should be given. When drafting the report it will also be necessary to explain any technical terms that have been used, as the client is unlikely to be familiar with them. For example, it will be desirable to explain the nature and purpose of the missing weatherboard. On occasions, it is necessary to make assumptions or possibly even intelligent guesses as to the reasons for failure, as in the case of the defective concrete step. Sometimes a defect can arise from a combination of factors. For instance, the poor condition of the concrete step could have been caused by a weak concrete mix, dirty aggregate, poor quality cement, insufficiently mixed concrete with the wrong water content and/or insufficient protection of the newly laid concrete against frost.

Drainage

The local authority has only a 6" (150) diameter foul sewer in Quebec Road, and will not permit the entry of surface water into this small diameter sewer which is laid to a fairly flat gradient. Hence the foul drainage is taken to the public sewer and the surface water from the roofs of the house and the garages is taken to soakaways.

House drainage systems will vary according to the type of sewerage system available, which may be combined, separate or partially separate and this needs establishing at the outset. In rural areas there may be no main drainage available, and the house discharge may be taken to a

Foul drainage

The main foul drains are laid in two lengths – the upper length runs from a manhole at the rear south-west corner of the house to the front south-east corner, and the lower length runs from the second manhole to a manhole on the public sewer under the public footpath in front of the house. There is a connection to the upper manhole, taking the discharge from an open gully serving the kitchen sink. A connection to the lower manhole takes the discharge from a 4″ (100) diameter cast iron soil and vent pipe which takes the discharges from the bath, washbasin and two WCs.

The drains are in 4″ (100) diameter clay pipes laid in straight lines between manholes to a gradient of 1 in 40, to provide a self-cleansing velocity. The pipes are jointed in cement mortar.

All foul drains were tested with a water test to a head of 3 ft (900) at the upper end of each length for a 30-minute period. This is a rather severe test for existing pipes but it will identify any major weaknesses in the drains. The tests showed a significant leakage in the length of drain between the two manholes and this drain is located only 1′-6″ (450) from the flank wall of the house and must be very close to and below the house foundations. A mirror test also showed irregularities.

It is likely that some of the drainpipes have been subjected to considerable pressure and the rigid cement mortar joints provide no flexibility. It is reasonable to assume that some of the joints are leaking and this could adversely affect the house foundations, if left unattended, and cause settlement. It is strongly recommended that this drain be opened up in sections until all the defective parts have been located. In view of its close

septic tank installation or a cesspool.

The layout of the drains should be described and the materials and sizes of pipes given in the report. If practicable the gradients of the drains should be determined to try and establish whether they are laid to self-cleansing velocities and so likely to remain free from silting.

The location of drain runs is important as they can become defective if laid close to buildings below foundation level, or in the vicinity of large trees and shrubs. Most drains to older properties consist of clay pipes jointed in cement mortar, often with a very strong mix such as 1:1. A mix of this type will shrink on setting and the mortar is likely to crack soon after the pipes are laid. It does not therefore provide a satisfactory method of ensuring watertight drains. Most modern properties are served by clay drains with flexible joints which are far more satisfactory. Where pipes pass under buildings it is advisable to use cast iron pipes and to provide means of access at each end.

The watertightness of drains should be tested using one of the recognised methods, such as a water test or air test. Tests for line, uniform gradient and freedom from obstruction can be carried out using a mirror at one end of the drain and a lamp at the other.

proximity to the house foundations, it would be good practice to replace the entire length – 24'-0" (7.315) – with cast iron pipes with flexible joints, to prevent any future problems, although this is an expensive item and must be carried out with care.

Manholes

The manholes are in one-brick walls in red engineering bricks laid in English bond in cement mortar and rendered internally. The upper manhole is 4'-0" (1.200) deep and the lower manhole is 4'-6" (1.350) deep. Each have half-round glazed clay channels and the benching and connections are well formed. There are however no step irons and access to the drains for inspection or rodding is therefore difficult. The fixing of three step irons spaced 12" (300) apart both horizontally and vertically would be a considerable advantage.

Both manholes have internal dimensions on plan of 3'-0" x 2'-3" (900 x 675) which is cramped. Each manhole is covered with a 6" (150) deep reinforced concrete slab which supports a 24" x 18" (600 x 450) cast iron light duty manhole cover and frame.

The rendering is breaking away from the internal faces of the manhole walls and could cause a blockage in the drains. The rendering is unnecessary as dense waterproof engineering bricks have been used in the manhole walls. It is recommended that the rendering to both manholes be removed and the brick joints flushed up with cement mortar.

One of the manhole covers is cracked and in dangerous condition and needs replacing, and both frames need rebedding in cement mortar and the covers bedding in grease to make them airtight.

In very old properties the pipe joints may be formed in puddled clay and they are unlikely to be watertight. This can be a very serious matter if the drains pass near wells from which the water is used for domestic purposes.

Full particulars should be noted in the survey of all manholes, inspection chambers and other access points. The form of construction and dimensions of manholes are important as well as a note of their condition.

Where chambers do not exceed 3'-0" (900) in depth, half-brick walls are normally adequate. Deeper manholes need one-brick walls and an increase in the dimensions on plan to provide adequate space for inspection and rodding purposes.

On cheaper developments, the manholes are often constructed in common bricks, rendered internally with a cement and sand mix. The conditions in a manhole are humid and aggressive and often cause extensive deterioration of the rendering and the bricks. Hence it is better to use fair faced engineering bricks.

In deeper manholes the top of the benching should not exceed a slope of about 1 in 6, so that a man can stand on it. All connections should point in the direction of flow. The type and weight of cover is determined by the type of traffic that it will have to carry. For example, lightweight covers can be used in flower beds where they will not be subject to vehicular traffic.

Ancillary features

The connection to the public sewer through a manhole is well formed and flows freely in the direction of flow.

The gully to the kitchen sink is silted up and the trap needs emptying, and the grating is badly corroded and needs replacement.

The 4″ (100) soil and ventilating pipe is in cast iron with caulked lead joints and is located inside the house, with an extension passing through the roof covering with a suitable watertight joint and a copper wire balloon to prevent birds nesting in the top of the pipe. The pipe was tested with a smoke test and was found to be satisfactory.

With old properties, it is customary to find intercepting traps and fresh air inlets. The intercepting traps are often blocked and the fresh air inlets damaged and badly sited.

Particulars of soil and ventilating pipes should be recorded, along with gullies and other fittings. Any tests and their results should also be included. Modern properties may incorporate the single stack system.

Surface water drainage

The rainwater downpipes to the house and garages discharge over gullies connected by clay pipes to soakaways. One soakaway serves the two downpipes at the front of the house, another serves the two downpipes at the rear of the house, and the third soakaway takes the rainwater from the garages.

The first two soakaways appear to be about 4′-6″ (1.350) diameter and the third about 3′-0″ (900) diameter. Each soakaway was planned to have a 3′-0″ (900) depth of hardcore below the drain inlet. Judging by the wet condition of the ground around each soakaway, it is probable that the soakaways are of insufficient size to cope with maximum flows in the highly impermeable heavy clay subsoil. It is recommended that the soakaways be opened up for inspection and, if necessary, that they be enlarged. Each of the soakaways is located in a grassed area of the plot.

The type of surface water disposal system should be described and its effectiveness examined. The use of soakaways in heavy clay soils is always suspect, and it is far better to take the discharges of rainwater into sewers or watercourses if at all practicable. In the case of the property being surveyed, there was obviously no alternative and, if the soakaways prove to be inadequate, the only available remedy is to enlarge them and backfill the excavated area with good, hard granular fill.

It is often good practice to provide an annotated layout plan of the drainage system.

External Works and Adjoining Properties

Paved areas

A 3'-0" (900) wide concrete path connects the front gate, front and back house doors, paved patio and garages. There are three defective areas, each of about 4 sq yds (4 m²) which are disintegrating and need replacing. There are several shrinkage cracks across the paths owing to the lack of expansion joints, but these do not merit any specific action.

A paved patio adjoins the rear walls of the lounge and dining room, 24'-0" x 5'-6" (7.315 x 1.676), consisting of coloured pre-cast concrete paving slabs bedded in lime mortar. Twelve of the slabs have settled badly and need relaying.

A concrete apron connects the garages to the rear service road and this is in satisfactory condition.

The form of construction and dimensions of paved areas in drives, paths, terraces and related areas all need recording, together with information about their condition and any remedial works considered necessary. On occasions, paving in large areas requires laying to falls to gullies, to assist with the disposal of surface water runoff. Tarmacadam and asphalt paved areas require resurfacing periodically.

Landscaping

The front and rear gardens are attractively landscaped and have been well maintained. Most of the area is grassed with good quality grass free from ryegrass and weeds. Groups of small flowering trees, evergreen shrubs, dwarf conifers and heathers together give an excellent blend of different shapes and colours.

A general description of the landscaping and the standard of maintenance should be included in the report. With large older properties the landscaping can be extensive and can form a special feature of the property, particularly when combined with water features.

Boundaries

The rear and side boundaries consist of plastic-coated chain link fencing, 3'-0" (900) high, attached to two straining wires, and 5" x 5" (125 x 125) reinforced concrete straining and intermediate posts at 10 ft (3.000) centres. The fence on the north-west side of the plot is in the ownership of the adjoining property. All the fences are in good condition, except for a 10 ft (3.000) length of the rear fence which is badly buckled and needs replacing. A beech hedge has been planted inside the chain link fence on all three boundaries and this is now well

Information concerning the boundary walls, fences and hedges is required, and details of any remedial works that are needed. Tall brick and stone walls in a dilapidated and unsafe condition can prove very costly to replace. With older properties, stone and brick retaining walls may be liable to collapse owing to deterioration of the stonework or brickwork, and sometimes there are no weep holes at

established and forms an excellent screen, providing ample privacy.

The front boundary wall is a one-brick wall in faced brickwork to match the house and has an average height above public footpath level of 2'-9" (825). The pressure of the earth behind the wall has caused a fracture in the centre of the length. The defective brickwork needs cutting out and replacing and it would be advisable to insert a brick pier at this point. The weep holes are silted up and need cleaning out.

A single oak gate, 3'-0" (900) wide, provides access to the property. The gate is soundly constructed with tenoned and pinned joints and is in reasonable condition, but it needs varnishing or staining and the defective fastener replaced.

the base of the wall. Front boundary walls to modern houses are often built in half-brick walls in porous bricks which are unsuitable for the conditions which they will have to withstand.

Gates can be either single or double, take various forms, be constructed of either wood or metal and should preferably conform to the appropriate British Standards. Rails should desirably be through-tenoned into stiles and each tenon pinned with a hardwood pin or non-ferrous metal star dowel. Metal gates may be constructed of wrought iron or mild steel and should preferably have a continuous framework, and be truly square and welded at junctions.

Garages

Two single garages are built side by side at the southern corner of the plot with access by a concrete apron on to the local authority's concrete service road at the rear of the plot. The first garage was built in 1959 and the second in 1974 to match the original. The internal dimensions of each garage are 20'-0" x 9'-3" (6.096 x 2.819) and they are constructed of half-brick walls in stretcher bond in the same facing bricks as the house with 12" x 9" (300 x 215) piers beside the doors and 9" x 9" (215 x 215) piers at the back corners and in the middle of each side wall.

There is a bad and progressive settlement crack on the south-east side of the original garage, which will involve strengthening the foundation and cutting out and making good the defective brickwork.

The floors are constructed in concrete and the roofs consist of 1½" x 7" (38 x 175)

The garages need to be just as carefully surveyed as the house and this would also apply to any other outbuildings which might be present. The client will need to know the form of construction and internal dimensions, and also the condition of the buildings and the nature of any necessary works of repair and replacement.

In this case the original garage has been extended by the addition of another single garage, which is not uncommon as a considerable proportion of households now have two cars and would prefer a double garage.

There are a number of structural defects which will be quite

roof joists at 18″ (450) centres, supporting corrugated asbestos sheets which fall to the fronts of the garages. A 1″ × 7″ (25 × 175) painted fascia board surrounds the garages at roof joist level, and needs repainting. The asbestos cement sheeting is laid to a shallow fall and there is evidence of some rainwater penetration at the joints in heavy storms. The sheeting to the roof of the original garage has weathered badly and is becoming very brittle and should be replaced. At the same time it would be advisable to consider reconstructing the roofs to both garages to a minimum pitch of 12° or to substitute the asbestos cement sheeting with boarding and bitumen felt.

The roofs drain into a 4″ (100) cast iron half-round gutter discharging into a 2½″ (63) cast iron downpipe and thence into a gully. The rainwater goods are rusting badly and need thorough cleaning and repainting.

Each garage has a metal window with a reinforced concrete lintel over the opening and a weathered and throated wood sill. The windows have been neglected and urgently require painting.

The original garage has a pair of matchboarded, framed, ledged and braced doors, 8′-6″ × 7′-6″ (2.690 × 2.286) overall, hung from a 4½″ × 3″ (112 × 75) frame. The frame is sound but the doors are in an advanced state of decay and require replacement. The second garage has an aluminium up-and-over door which is in good condition apart from the paintwork. It would probably be good policy to replace the defective pair of wooden doors to the first garage with a single aluminium up-and-over door to match the one in the newer garage.

Adjoining properties

There is a small detached house on the south-east side of the property and a pair of semi-detached police houses on the north-west side. Both properties were built at about the

costly to rectify. Hence the surveyor when carrying out the survey must visualise all the remedial works that will be needed and be able to estimate the probable costs involved. In cases where alternative solutions are available, the options must be described, and this applies in the case of the roofs and the replacement doors to the original garage. All significant dimensions and sizes of components should be recorded and it is always better to have too much information than insufficient, as return visits to properties can be expensive.

In the case of older properties, there may be stables, conservatories and a range of outbuildings, all of which require careful scrutiny. The surveyor cannot be too careful and must continually be on the lookout for defects of all kinds. He cannot take anything for granted and he carries a heavy professional responsibility.

Brief details should be given of adjoining properties and of any restrictive covenants or easements which could adversely

same time as the property being surveyed and are built to the same building line. There are no unusual restrictive covenants or easements, and nothing which should interfere with the use and enjoyment of the property. affect the use of the property, such as rights of way, light or drainage.

Report to Client

The survey will be written up in the form of a report, following the sequence and headings adopted for the survey. The report should contain a summary of the main findings and a schedule of defects with the estimated costs of rectifying them. The client will then be fully aware of his probable financial commitments if he wishes to purchase the property. He will also be well informed on the standard of construction of the property which will help him in coming to a decision. A report of commercial/industrial buildings is contained in chapter 5, accompanied by explanatory notes.

5 Structural Reports

This chapter details the essentials of good report writing and then illustrates their application to specific buildings, accompanied by guidance notes, primarily to help students.

Essentials of Good Report Writing

Purpose and Nature of a Report

A report is a written document produced as a result of investigations and/or research undertaken to reveal information.[1] Its primary purpose is to give information which is of specific interest to the client, and the report must meet the client's needs. There may be instances where the client is not wholly sure of what he wants, and it will then be necessary for the surveyor or other professional preparing the report to liaise closely with the client and it may be advisable for him to compile the brief for approval by the client.

In practice the surveyor's instructions will probably originate from an interview or letter which will describe the purpose of the report and any particular matters which require special attention. In an examination, a question will replace the interview. In order to be clear on the purpose of the report and the matters to be considered, the student should read the question several times. It is good practice to make rough notes of the relevant information each time the question is read and to underline significant phrases in the question for emphasis. The student is then in a position to produce a framework of his report planned in a logical sequence, on the basis of which the report can be drafted.

In practice the methods employed by the surveyor in the preparation of the report will be influenced considerably by his own special knowledge and experience and the terms of reference. Familiarity with similar situations in the past will assist him in determining his approach to the writing of the report. A unique building may call for a special approach.

A RICS Guidance Note[2] emphasises that the diversity of properties and client requirements demand a flexibility of approach to ensure that relevant information is provided in the most concise and clear manner. It is vital that the surveyor should not lose sight of his objectives which are primarily to give his findings clearly, definitively and unambiguously. He should always distinguish clearly between fact and opinion and may benefit from the continual use of a particular format for all reports. It is imperative that no statements should be made or opinions given which indicate a level of knowledge beyond the surveyor's capabilities.

Content of a Report

The general approach outlined by Cooper[3] in the preparation of draft technical reports is applicable to structural reports. He identifies the main stages as: collection of material; selection; logical ordering; interpretation; and presentation.

Darbyshire[1] identifies three characteristics which distinguish reports from many other kinds of written documents.

(1) The most important is objectivity. The report writer's main task is to set down the relevant facts, without regard to his personal feelings or views about them. He often makes recommendations which must be adequately supported by the facts. The report is an important document which provides evidence on which decisions will be made and action taken by the client.

(2) A report which is well prepared and produced will generally have a distinctive form. The arrangement of the parts, as described later, will clearly identify the document as a report.

(3) A report generally has a characteristic style, although the actual style will vary with the writer. A suitable style normally follows naturally from the correct use of objectivity and form. Darbyshire[1] describes how style is a function of language use. In a report, specific facts have to be stated within the restrictions imposed by its form, and the style adopted should be the most suitable for the expression of fact. The writer must observe good grammar and the principles of sentence structure and aim at good writing.

A good report normally follows a logical approach, which is framed to answer the specific points, requests and enquiries made by the client. The contents will depend on the purpose for which the report is required. A common approach is now outlined.

(1) *Introduction.* The report normally starts with a suitable heading which should be kept clear and concise. This is often followed by a brief summary of the instructions given by the client which constitute the terms of reference and the purpose of the report, the date the report was written and by whom.

(2) *Recital of the facts.* This will form the main body of the report and will largely be the result of the surveyor's inspection of the property. Facts can be subdivided into two categories, namely descriptive facts and non-descriptive facts.

Descriptive facts contain information obtained from an inspection of the property and may include such items as the situation of the property, type of structure, sizes of rooms, structural defects and the like. Non-descriptive facts comprise information which cannot be obtained by observation and may include such matters as tenure, rating assessment, easements and similar aspects. One of the greatest problems in presenting facts is to avoid over-complexity, and the writer needs to keep a clear mind with a singleness of purpose. The essential characteristics are clarity, conciseness and accuracy.

(3) *Conclusions.* The report normally finishes by summarising the principal matters in the body of the report and presenting conclusions. Advice should be given on all matters on which the client requires guidance to enable him to make the appropriate decisions.

A RICS Guidance Note[2] contains some useful guidelines on the content of structural reports. It emphasises that the report should follow a clear and logical sequence and yet remain readable. Flexibility of approach is important as the sequence of reporting on one building or for one client may be irrelevant in another case. For example, it may be appropriate to relate the report to elements of construction in one instance, whereas sections of the building or separate tenancies may be more relevant in another case.

In most buildings, and particularly the more complicated ones, the identification of the different parts of the property is probably best achieved by floor plans, possibly in the form of line drawings, as illustrated in figure 4.

In order that the report shall read well and relate principally to essential facts and advice, much of the more precise detail can with advantage be incorporated in appendixes. Appendixes could, for example, contain schedules of defects, details of tenancies and relevant repairing covenants, and consultants' reports. The indiscriminate use of caveats should be avoided but insurers may require the inclusion of certain clauses and these should be stated *verbatim.* To reduce the possibility of claims by third parties, it is common practice to include a suitable clause, which could take the following form. 'This report shall be for the private and confidential use of the clients for whom the report is undertaken and shall not be reproduced in whole or in part or relied upon by third parties for use without the express authority of the surveyor.'[4] It may also be necessary to make reference to the inaccessibility for any reason of particular parts of the property.

The report normally contains a summary and firm recommendations in respect of the property. A RICS Practice Note[4] relating to structural surveys of residential properties recommends that the summary to a report should:

(1) provide a broad assessment of the construction and condition of the property relative to other similar properties;
(2) advise the client to consider the report as a whole rather than to take out of context conditions of disrepair, which may be normal for the type and age;

(3) emphasise any serious reservation, defect or condition; and

(4) offer to discuss and advise on any points of difficulty arising out of the report.

The position within the report of the summary of the main points and the recommendations can vary from inserting them at the start of the report to including them at the end. Some clients prefer to have them at the beginning so that they can immediately obtain an overview of what the report contains. It will certainly help the reader who is short of time by informing him what to expect in the main body of the report. The main argument against this arrangement is that the reader may be inclined to pay less attention to the main body of the report and thus overlook many important points.[5]

Use of Covering Letters

A report is often accompanied by a short covering letter to the client. The letter could make reference to the client's instructions, the inspection of the property, the enclosing of the report and any incidental matters which the surveyor considers should be drawn to the attention of the client.

The student should bear in mind certain basic principles in the drafting of such letters:

(1) if the letter begins with Dear Sir, it should end with Yours faithfully;

(2) if it begins with Dear Mr(s), it should end with Yours sincerely;

(3) a check should be made to ensure that the letter is dated;

(4) both the client's and surveyor's references should be inserted; and

(5) a check should be made to ensure that all appropriate enclosures are enclosed.

Presentation

A RICS Guidance Note[2] rightly recommends that a report should at least be broken up by prominent headings and provided, where appropriate, with a logically numbered index to allow speed of reference. In all cases the report should be well presented and carefully checked to ensure that the final document brings credit to the surveyor or other professional preparing the report.

Gass[6] postulates that in order to satisfy the client, a report should:

(1) be well presented;

(2) be clear and well written;

(3) arrive at definite conclusions;

(4) often include a valuation;

(5) provide accurate, intelligent and practical information; and

(6) be produced with speed and efficiency.

Essentials of a Good Report

Certain well-established criteria should be observed in writing structural reports, and these will each be considered.

Neatness. Even the best prepared and most accurate reports will suffer from being produced in an untidy manner. In practice, reports will be typed, preferably from a well-prepared manuscript or carefully dictated notes or tape. In an examination, reliance will be placed on handwriting which should be clear and legible. *Use of English.* Reports should be prepared in simple, direct, grammatical English and be free from spelling mistakes. The quality of the thought conveyed, the clarity of the views expressed and the understanding by the reader of the contents of the report all depend on the correct use of English. Most bad English emanates from a lack of clear thinking.

Technical writers often tend to make their sentences too long and frequently display an excess of verbosity and lack of conciseness. Darbyshire[1] believes that easily read sentences often do not exceed ten to fifteen words, although sentences of up to twenty five words can normally be understood, provided that they are well worded. Sentences of more than thirty words should, in general, be avoided, although this is not always the case in practice.

A good writer will not use long words where short ones are available. A high proportion of long words in a sentence can make it difficult to understand. Always endeavour to obtain the best word to fit the particular case.

Spelling is important as bad spelling brings discredit on the writer. All surveyors and students should have a copy of the *Oxford English Dictionary* at hand for reference when in doubt. As a general policy it is advisable to avoid abbreviations as far as possible.

For numbers below ten it is normal practice to use words rather than numerals. Avoid using numerals to begin a sentence, and do not use two sets of numerals in succession as, for example, 12 25 mm bolts.

Hyphens and quotation marks are generally better avoided. Punctuation helps the writer to make his meaning clear, and enables the reader to read quickly and without ambiguity. Commas provide a momentary resting place for the eye and their correct use, if there is such a thing, is acquired by common sense, observation and careful thought. Some writers use the semi-colon sparingly and the colon not at all, and others use dashes. Cooper[3] considers that the main problem facing technical writers is not of knowing whether or not to use a colon or semi-colon, but of breaking up over-long sentences into smaller and more meaningful units. Care should also be taken to avoid the use of excessively long paragraphs. *Style.* A terse style is advisable in report writing in order to convey the most information in the fewest words. Phrases such as 'It must always be remembered', 'on the other hand' and 'in this connection' are best avoided. The writer should avoid all forms of padding, such as repetitions for effect, irrelevancies and verbosity generally.

Accuracy. It is most important that a report shall be accurate, as the existence of errors or vague statements will detract considerably from the value and credibility of the report.

Simplicity. The majority of structural reports are prepared for the benefit of lay clients and so they should be written to be readily understood by persons without technical knowledge. Where technical points are included they should be explained clearly.

Clarity. The report must be easily readable, prepared in a logical sequence with the various sections prefaced by suitable headings and sub-headings, to act as signposts and guide the reader to the required section.

Some surveyors use numbers to help identify the different sections of a report. For example, the initial numbering of the main elements of the report might be in Roman numerals (I, II, III, IV), with the first sub-divisions in Arabic numerals (1, 2, 3, 4), then small letters (a, b, c, d) and finally Roman numerals in lower case (i, ii, iii, iv). Alternatively the decimal system may be used starting with 1, numbering succeeding sub-paragraphs 1.1 and 1.2, and sub-sub-paragraphs 1.1.1 and 1.1.2. Sometimes only sections are numbered in this way, while in other instances each paragraph is numbered.[3]

Many surveyors regard extensive numbering as over-elaborate and unnecessary. Certainly it is necessary to question the value of such numbering and the extent to which it will help the reader to find his way through the report.

Continuity. The presentation and arrangement of information should be in a logical order in a series of paragraphs. Each paragraph should be about one topic, be complete in itself and yet so related as to lead to a conclusion through a series of steps. Although the facts may be presented as a number of complete paragraphs they should be so related as to be capable of being read as a complete narrative.

Variations of emphasis. The report should apply variations of emphasis to matters of varying significance to provide clients with a better understanding of the contents of the report.

Conclusions and Recommendations

The conclusions can summarise the findings and inferences, but should not contain any matters not previously introduced into the report. The surveyor must take care to distinguish between indisputable facts, the surveyor's firm opinion, and any necessary speculative comments included for guidance or prognosis.[4]

Although recommendations appear separately from the conclusions in some reports and are bracketed together in others, logically they should be linked, as they form part of the conclusions. Recommendations are often the most difficult part of the report to write and the surveyor must always consider carefully his relationship with the client in formulating the recommendations.[3]

Production of the Report

Many surveyors produce their reports with a standard cover and backing sheet so that all their reports are immediately recognisable. This helps to give a good impression and also provides the opportunity to have the name of the practice or organisation, address and telephone number pre-printed on the cover with the word REPORT included in large letters.

The client's name, the subject of the report and the surveyor's name and the date should be given on the title sheet, and this may be followed by a contents sheet. It is good practice to start each fresh section on a new page unless the report is very short.

Chappell[5] has provided some useful guidelines for the production of technical reports which deserve careful consideration.

(1) Leave generous margins, particularly adjoining the spline. This helps with binding and the client will find the space useful for pencilling in any comments.

(2) The inclusion of adequate headings, sub-headings, and the numbering of pages and possibly sections and paragraphs will help the reader in assimilating the report. Some surveyors prefer to have the address and reference of the property surveyed inserted on each page.

(3) Some surveyors have the type double spaced, but this is not absolutely necessary if adequate margins are provided. The combination of double spacing and generous margins tends to produce an unnecessarily bulky report with a limited amount of information on each page.

(4) The inclusion of notes and references is best avoided, as they interrupt the flow of the text and the reader may find them irritating. If they are considered vital, they can be inserted either as footnotes, sidenotes in the margins or endnotes.

(5) Any additional information, such as drawings, photographs, tables and statistics, is best inserted in appendixes.

Reports should be typed with modern typewriters using a suitable typeface on good quality paper. The report should be properly bound so that it can be opened flat; a variety of different forms of binding is available as shown in table 5.1.

Report on Transport Depot

This report covers industrial/commercial buildings, paved areas and drainage, and is considered to have sufficient scope to be of value to practitioners and students alike. The same approach has been adopted as with the survey in chapter 4 by adding explanatory notes beside the report to increase its usefulness and enlarge on important points.

Table 5.1 Principal methods of binding reports

Method of binding	Advantages	Disadvantages
Plastic spiral	Opens flat, relatively cheap and good appearance	Can tear the paper, requires special machine and generous margins and can become tangled
Plastic clip spine	Cheap and easy to apply	Difficult to open flat and pages become detached
Folding staples and bar	Cheap and easy to apply	Difficult to open flat, untidy appearance and steel can cause cuts
Staple, washer and single hole	Cheap and easy to apply	Difficult to open flat and unsuitable for thick reports
Ring binder	Good appearance, opens flat, relatively cheap and easy to apply	Requires generous margins and can tear the paper
Glued spine and plastic binding	Good appearance, margins not critical and opens flat	Not cheap and pages can detach with use
Fully sewn, glued and bound	Good appearance and margins not critical	Expensive and requires professional techniques

Source: D. Chappell, Report Writing for Architects[5].

REPORT
by J. P. Charterhouse FRICS
on
Transport Depot off
Sloane Street, Newton Bassett,
Northamptonshire
for
Midlands Transport Services Ltd.

I General Particulars

The site is located at the rear of 23 Sloane Street, and was originally part of the garden of this property. It occupies approximately half an acre (0.20 ha). It is approached from Sloane Street by a communal tarmacadam

The heading to a report can take a variety of forms and will often be contained on a separate title sheet. The report may also be accompanied by a covering letter which outlines the purpose and date of the survey, the client's instructions and other relevant factors. The property must be identified and also the client for whom the report is being prepared.

A general description of the site and the buildings will assist the client in obtaining an overall picture of the property. This particular property is situated on access way 1360 ft (415 m) long and has an average width of 19 ft (5.8 m) This access way leads on to a concrete access road, in the ownership of the property being surveyed, 1900 ft (580 m) long and 10 ft (3 m) wide, with an intermediate waiting bay.

The three blocks of buildings surround a concrete paved area of approximately 400 sq yds (335 m²). The site is bounded on the east side by a hedge and a painted cast iron fence – 6 ft (1.8 m) high – and on the other three sides by one-brick walls varying in height from 9'-0" to 12'-6" (2.74 to 3.81 m).

The three buildings on the site consist of two garage blocks, 60'-0" × 25'-1½" × 11'-0" headroom (18.288 × 7.624 × 3.353) and 32'-0" × 19'-4" × 8'-6" headroom (9.754 × 5.892 × 2.690) respectively. The mess block is 32'-0" (average) × 14'-0" × 8'-0" (9.754 × 4.267 × 2.438) and contains toilets, ablutions, mess room and drying room. The layout is shown in figure 5.

The buildings were erected in 1960 by the former Newton Bassett Rural District Council as a central depot, principally to house six refuse collection and cesspool emptying vehicles and four vans. In 1974 the Rural District Council was absorbed into the newly formed Ousebury District Council and the property has now become surplus to the Council's requirements.

The paved area serving the buildings is over 8 ft (2.44 m) above normal river level and the highest recorded flood level is 1'-0" (300) below the lowest floor level.

The gross rateable value has been assessed at £1325, and the tenure is freehold.

back land and details of the access way are of considerable importance.

The general nature of the boundaries need describing and the nature and dimensions of the buildings are also included.

The date of construction of the buildings and their general history are significant. Their original construction by the former Newton Bassett Rural District Council can be assumed to ensure compliance with the Building Regulations. The capacity of the garages is also useful from the prospective purchaser's viewpoint.

A diagrammatic layout of the site is always useful.

Other matters of significance such as liability to flooding, adverse restrictive covenants, substantial structural repairs, tenure and gross rateable value should also be included at this stage.

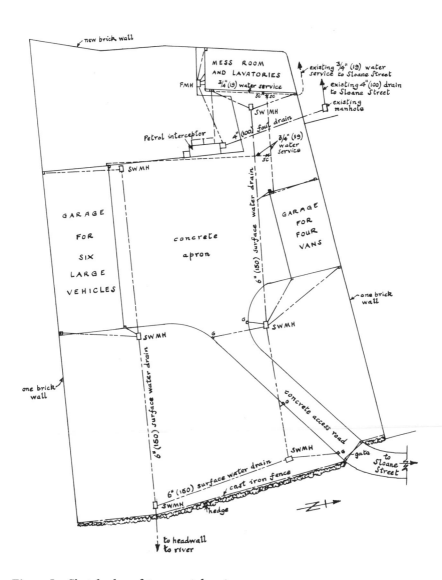

Figure 5. Sketch plan of transport depot

II Large Garage Block

2.1 Brickwork

The one-brick boundary wall on the south side of the site was rebuilt in 1960 in class B engineering bricks in cement mortar in English bond, with $18'' \times 18''$ (450 x 450) piers at $10'\text{-}0''$ (3.00) centres. Part of the wall forms the rear wall of the large garage and is finished fair face. The wall is in good condition with a satisfactory bitumen felt damp-proof course.

The two end walls are also in one brick walls in class B engineering bricks strengthened by a pier at each end. About 5 sq yds (5 m^2) of pointing is required to the external face of one end wall.

There are two hot dipped galvanised steel windows, each $5'\text{-}0'' \times 4'\text{-}6''$ (1.524 x 1.372), with an opening light, in each end wall. These have been neglected and are rusting badly with eight cracked panes of glass and they require cleaning down, broken glass replacing and windows repainting.

The window heads consist of $4\frac{1}{2}'' \times 6''$ (112 x 150) reinforced concrete lintels on the inside face and brick-on-end arches supported by $3'' \times 3''$ (75 x 75) mild steel angles. The bottom legs of the angles are rusting and need cleaning down and priming.

The external window sills are formed of two courses of clay roofing tiles, a number of which are cracked and need replacing to prevent damp penetration. The mastic pointing around the windows has perished and needs replacing to make them watertight. The internal sills consist of clay quarry tiles and are in good condition.

2.2 Roof

The mild steel roof trusses are well supported by $7'' \times 4''$ (175 x 100) rolled steel stan-

The best approach to the report in this case is to take each building separately because of their diverse components and dimensions. In each case the construction needs to be described in detail, together with the defects and the best method of remedying them. The full range of defects will be listed in a later section of the report, together with an estimate of the cost of the remedial work, in order that the client can take these costs into account when deciding how much he can afford to pay for the property, after being satisfied that it meets his needs.

It could be considered that the report contains too much detail and too many technical terms. The omission of this information would however divorce the report of much of its value and reduce it to vague and superficial terms. The surveyor will be at hand to explain, or clarify, any points which the client is unable to understand fully.

A logical sequence is followed to avoid omitting important components. With this particular building the order of approach is brickwork, windows and openings, roofwork, rainwater goods, floor and inspection pit and doors.

With the inspection pit, it would be so easy merely to say

chions on the front face and are rag-bolted to concrete pads set on brick piers at $10'-0''$ (3.000) centres on the rear face. The roof trusses are well constructed with $2\frac{1}{2}'' \times 2'' \times \frac{1}{4}''$ (63 × 50 × 6) rafters and the remaining members are $1\frac{3}{4}'' \times 1\frac{3}{4}'' \times \frac{1}{4}''$ (42 × 42 × 6) angles with riveted joints. Eight lines of $3'' \times 2'' \times \frac{1}{4}''$ (75 × 50 × 6) purlins support the asbestos cement roof sheeting and the ridge. All the steelwork needs cleaning down, priming and painting for protection. Six of the asbestos cement sheets are cracked and need replacing and it is likely that all the roofing sheets will require replacement within a few years as they are becoming brittle.

2.3 Rainwater Goods

The metal box gutters at the rear and the asbestos cement eaves gutter at the front are of adequate capacity and are laid to satisfactory falls. The rear gutter is badly silted up and requires cleaning to prevent it overflowing into the building in times of storm. The $6''$ (150) diameter asbestos cement eaves gutter and the four $3''$ (75) diameter downpipes are all in very poor condition and should desirably be replaced with metal products, preferably aluminium for durability. The replacement cost could however be halved by using PVC rainwater goods.

2.4 Inspection Pit

The concrete floor is in sound condition apart from two minor cracks. There is a useful inspection pit $16'-6'' \times 2'-6'' \times 5'-0''$ (5.030 × 760 × 1.524) internally, constructed of one-brick walls in class B engineering bricks in cement mortar, supported on a concrete base and covered with $7'' \times 2''$ (175 × 50) loose boards.

that there is a satisfactory inspection pit of brick walls on a concrete base covered with loose boards. The client is, however, entitled to more information than this.

With regard to the roof trusses, some surveyors would argue that it is necessary to state only that these are steel roof trusses and to comment on their adequacy and condition. The inclusion of the sizes of the truss members shows that a thorough examination has taken place and provides the client with full details of the construction for future reference, should he subsequently purchase the property.

A similar aspect arises with the large sliding doors, where a detailed description of the rather outdated construction is given. A briefer description, omitting the sizes of the various members could easily be produced, but the author sees merit in giving a reasonably full acount of the construction, particularly as it is a problem area.

Where components are defective and require replacment, considerable thought must always be given to the available alternatives. The products available when the building was originally constructed are likely to have been superseded by more effective components and materials in more recent times. It is not therefore always simply a matter of replacing like with like, and the client will need advice on these aspects.

2.5 Doors

There are twelve timber sliding doors each 5'-3" × 10'-3¼" (1.600 × 3.874), each made up of 1⅞" × 9" (47 × 225) top and bottom rails, two 1¼" × 9" (32 × 225) intermediate rails, 4½" × 1⅞" (114 × 48) stiles and ⅝" (16) tongued and grooved, V-jointed boarding. The doors are suspended from effective sliding headgear bolted through a timber fascia to a steel channel which is bolted to steel brackets top and bottom at 4'-0" (1.200) centres. An inverted steel channel is inset into the base of each door which slides over a bulb tee set in the concrete floor.

The doors are showing signs of extensive wear, some of the joints between members are broken and the boarding is deteriorating with broken joints and cracked and rotten boards. It would be advisable to replace all the doors, and aluminium roller shutter doors operated by endless hand chains could provide a good, although expensive alternative, with considerable advantages in both operation and maintenance over the present type of doors.

2.6 Electrical Installation

There is no electrical power or lighting in the building.

III Small Garage Block

3.1 Brickwork

The rear wall consists of the original one-brick boundary wall heightened by about 2'-0" (600) and protected at the top by a weathered and twice throated precast concrete coping, Unfortunately the wall has settled badly in two places and it is still moving. The cracks indicate that the foun-

The next logical step is to follow with the small garage block, as it has some features in common with the large garage and some reference back will be possible to avoid repetition.

One major difference, apart

dations are insufficient to carry the brickwork, although no loads from the building are transmitted to it. The cracks could be filled with mortar and new bricks stitched into the brickwork as necessary, but in view of the size and location of the cracks, the wall cannot be considered structurally stable. It is advisable to underpin a 15'-0" (4.572) length of wall by excavating beside and below the existing foundation in 3'-0" (900) lengths and constructing a new foundation, wedged up to the underside of the existing. This can be done only from the adjoining garden, and the owner is prepared to allow this, subject to satisfactory reinstatement. It will, however, be a fairly costly item as shown in the cost of repair schedule later in the report.

Dampness has penetrated the wall up to a height of about 1'-6" (450), and there is no evidence of a damp-proof course. The most appropriate remedy is to apply a water-proofed cement mortar rendering or a proprietary damp-proofing material, such as 'synthaprufe', to a height of 3'-0" (900).

The two end walls are one-brick thick with a pier at each end and built in class 'B' engineering bricks in cement mortar in English bond. The brickwork is in sound condition, except for some repointing needed to the verges, and has an effective bitumen-felt damp-proof course.

3.2 Structural Steelwork

The roof loads are carried by an effective steel framework consisting of three pairs of 7" x 4" (175 x 100) rolled steel stanchions. The stanchions are given further support by 6" x 3" (150 x 75) rolled steel channels bolted to the tops of the front stanchions and 4" x 3" (100 x 75) rolled steel joists between the rear stanchions. All the steel-

from the smaller size, has been the decision, at the design stage, to use the existing boundary wall as part of the enclosing structure to the garage. Its condition at the time must have been considered suitable, although the designer was sufficiently prudent to include a rear line of stanchions to carry the roof load. The roof in itself may not be particularly heavy, but when covered with a thick layer of snow, the load will be considerable. However, in the event the foundations have not been able to cope with the weight of increased quantity of brickwork, and some failure has resulted.

The work of underpinning the existing foundation requires explaining to a lay client as the term will probably not mean very much to him. The difficulties of undertaking this work and the high cost involved require particular emphasis.

The absence of a damp-proof course in the boundary wall has given rise to dampness problems. The surveyor needs to explain this and to come forward with suitable remedies.

The surveyor must satisfy himself as to the adequacy of the roof structure and investigate its condition. The asbestos cement roof sheets are becoming very brittle after 24 years and the surveyor is wise to recommend their replacement in these circumstances. This is followed by consideration of the rain-water gutters and downpipes –

work is rusting badly and needs cleaning thoroughly, priming and painting.

The three steel roof trusses are effectively designed and pre-fabricated and apart from extensive rusting are well suited for their task of supporting the asbestos cement sheeted roof. The rafters are $2'' \times 2'' \times \frac{1}{4}''$ ($50 \times 50 \times 6$) mild steel angles and all the other members are $1\frac{3}{4}'' \times 1\frac{3}{4}'' \times \frac{1}{4}''$ ($42 \times 42 \times 6$). There are also eight lines of $3'' \times 2'' \times \frac{1}{4}''$ ($75 \times 50 \times 6$) angle purlins.

3.3 Roof Covering

The asbestos cement sheeted roof is in an advanced state of deterioration as the asbestos cement has become very brittle. The whole of the sheeting and the ridge capping should be replaced before it permits damp penetration.

3.4 Rainwater Goods

The rear metal box gutter is in good condition but needs clearing of accumulated silt. The lead flashing, which makes a watertight joint between the face of the brickwork and the gutter, has become distorted and needs pressing back into position and repointing. The front $6''$ (150) diameter half-round asbestos cement eaves gutter and the four $3''$ (75) diameter asbestos cement downpipes have effectively reached the ends of their respective lives and need replacing, preferably with metal products.

3.5 Doors

There are eight timber sliding doors $4'$-$3'' \times 7'$-$9\frac{1}{4}''$ (1.295×2.367) of similar construction to those in the large garage. Although they have not reached quite such an advanced stage of deterioration, they do nevertheless justify replacement by suitable metal doors

their adequacy and condition. Again the original choice of material was not the best and the designer should always have regard to future costs in the formulation of his design.

The sliding doors, although of different size, are of similar construction and condition to those in the large garage. Hence it is unnecessary to repeat the full constructional details and the suggested method of replacement.

In some cases it is necessary to spell out to the client the purpose of a particular building component, with which he is unlikely to be familiar. A typical example is the lead flashing where a suitable description is inserted.

The absence of electrical power and lighting in a building of this type is exceptional and deserves special mention, as the client may wish these services to be provided should he decide to purchase the property.

as, if left, they will be a constant source of trouble.

3.6 Concrete Floor

A small area of the concrete floor – about 24 sq ft (2.24 m²) – is spalling badly and needs cutting out and making good.

3.7 Electrical Installation

There is no electrical power or lighting in the building.

IV Mess Room and Lavatories

4.1 Accommodation

This building has access through an entrance lobby to two WCs, an ablutions room and a mess room. A drying room is entered from the mess room. The ablutions room is 8'-3" × 6'-6" (2.514 × 1.981), the mess room is L-shaped with overall dimensions of 14'-0" × 12'-6" (4.267 × 3.810), and the drying room is 14'-0" × 11'-6" (4.267 × 3.505), taking an average width. All the rooms have a height of 8'-0" (2.438).

4.2 Brickwork

The rear wall and one end wall consist of the original one-brick boundary wall, while the other two walls are one-brick walls in class B engineering bricks in cement mortar in English bond, all covered with a suitable precast concrete coping. All internal walls are half-brick walls. The new brickwork is in good condition, except that the bitumen-felt damp-proof course needs repointing.

There is an area of dampness of about 5 sq yds (5 m²) above the floor to the rear wall in the mess room. The mortar joints on

The final building to be covered in the report is the mess room and lavatories. It will assist the client if the individual rooms are described and their leading dimensions included. The room heights should be inserted in addition to the dimensions on plan. The construction of the structure should be described and its condition noted. It is important that mention is made of the boundary walls forming two of the enclosing walls of the block. Defects are more likely to arise on this part of the structure than in the newer sections and this has proved to be the case. It also highlights the dangers of using building elements in common with other owners, as the method of use of the other side of the element cannot always be controlled and unsatisfactory operations can go unnoticed.

the opposite side of the wall are open and garden soil has been piled up against the face of the wall by the adjoining property owner. It is recommended that the owner be approached with a view to making good the mortar joints and repositioning the soil.

4.3 Roof

The roof to the block is constructed of a concrete slab supported by external and internal walls and a boxed-in rolled steel joist running the length of the building. The concrete is covered with a screed and asphalt which is laid to a satisfactory fall and taken 6″ (150) up the parapet walls, and the top edge of the asphalt skirting is taken into a chase in the brickwork. In three places the asphalt upstand has cracked, allowing some rainwater to penetrate the junction of the wall and ceiling in the drying room and mess room. The sections of defective asphalt upstand require replacing.

The surface water from the roof discharges through two asbestos cement hopper heads into 4″ (100) diameter asbestos cement pipes which are in satisfactory condition.

Roofs always need careful scrutiny, particularly flat ones, as they can so often be the source of damp penetration. The asphalt surface may be split or cracked owing to movement of the structure. Cracked and blistered areas should be heated, cut out and made good with new asphalt.

4.4 Windows

There are galvanised steel windows, with pivoted opening lights to all the rooms. They are rusting badly and need thoroughly cleaning down, priming and painting. The construction of the sills and window heads is identical to that for the large garage blocks, and all are in sound condition.

The steel windows are made of sections about $\frac{1}{8}″$ (3) thick which, if these are permitted to rust extensively, can lead to considerable distortion of the windows and cracking of the glass. It is important to keep them relatively free of rust and regularly painted.

4.5 Doors

The external door is a 2′-9″ × 6′-8″ × 1¾″ (826 × 2040 × 44) standard softwood panelled door, and fixed to a 4″ × 3″ (100 × 75) frame, while the five internal doors are

The joinery is described with any defects noted and the remedial work included in the report. External joinery in timber con-

standard half-solid flush doors faced with plywood and hung from 1½″ (38) rebated linings. Two of the internal doors are badly damaged and require replacing and there are three sets of defective lever door handles.

taining excessive sapwood is constantly causing problems in modern buildings, as described in chapter 3. Attention should also be paid to ironmongery and its adequacy and condition.

4.6 Internal Finishes

The floor finish to all rooms is of granolithic concrete which is in acceptable condition despite the presence of some minor crazing. All the walls and ceilings are finished in gypsum plaster which is generally in fair condition, apart from the damp areas, previously described, which will need replacing after the damp penetration problem has been solved. The ceiling to the ablutions area is crazed and the cracks need cutting out and making good. A small area to one of the walls in the drying room − 3 sq yds (3 m²) − is loose and needs cutting out and replastering.

The walls and ceilings are all finished in matt oil paint which is very dirty and needs cleaning down and repainting, following the execution of the recommended repair work. The junction between the walls and floor is covered by a 1″ × 5″ (25 × 125) chamfered softwood skirting which is in good condition.

The report then proceeds logically from joinery to internal finishes. As is often the case, external defects will have their counterpart internally, and this applies particularly to settlement cracks and damp penetration. Plastering defects can take a number of different forms as described in chapter 3.

With all defects, the client will want to be informed of the remedial work that is needed to correct the defects, and a summary of the costs of the various repair items will be included later in the report.

4.7 Sanitary Appliances

The two WCs are white ceramic low down suites. One has a cracked pan and the other a defective ball valve, both of which need replacing. The three stall urinal is in fair condition, but the two wash basins in the ablutions section are badly crazed and do need replacing with more modern-type fittings. The stainless steel sink and drainer in the mess room is satisfactory, as are all the copper hot and cold water service pipes and waste pipes.

The report then proceeds to deal with the various sanitary appliances. They appear to be the original fittings installed in 1960 and are showing signs of wear and tear. If the number and type of fittings is unlikely to meet the needs of the client, this should be stated here or better still in the conclusions section of the report.

4.8 Electrical Installation

The electricity supply enters the block through the mess room which contains the meter and a switchfuse control unit. The cables are PVC insulated and sheathed and are in good condition. The system withstood satisfactorily the Electricity Board tests. There are ample lights and power socket outlets but these are deteriorating and should be replaced with modern fittings. The electric water heater over the sink and the electric convector heaters in the mess room and drying room have recently been replaced and are all in very good condition.

The form and condition of the electrical installation needs including in the report. With older installations, it is likely that rewiring will be needed and this will be a fairly costly item. Attention should be paid to all associated fittings. Electrical services often form a separate section in the report, but it is included with the mess room and lavatories in this case as there is no electrical supply elsewhere on the site.

4.9 Telephone

A Post Office telephone is located in the mess room.

The existence of the telephone is worthy of mention as an additional asset.

V Paving and Access Road

The access road in the site ownership is constructed of a 10'-0" (3.00) wide reinforced concrete road laid on hardcore with a cross-fall of 2½" (64). Expansion joints are provided at 60'-0" (18.288) centres, which need resealing. There is a waiting bay with an average length of 40'-0" (12.192) at about the half-way mark.

The length of road from the site boundary to the tarmacadam communal road, has a 12" (300) wide trench, adjoining the lower side of the road, filled with gravel rejects, for surface water drainage.

The road within the depot site has a 5" x 10" (125 x 250) precast concrete kerb on both edges, and surface water is discharged through gullies into the surface water drainage system.

The paved area between the buildings is also constructed of reinforced concrete laid

It seems logical to follow the buildings with the paving and access road, giving a description of the construction as far as practicable. There was no easy way of determining the depth of the concrete slab, but it is assumed to be 6" (150). The siting of the expansion joints indicates that the concrete is reinforced with steel fabric.

The kerbing and surface water drainage arrangements should also be described. Any defects are then listed together with the remedial works that are considered necessary.

Had the road been metalled, it is likely that it would have needed surface dressing or even

to falls to drain to gullies. It is provided with adequate expansion joints, although most of them need cleaning out and refilling with flexible joint filler.

Three bays of the concrete access road are badly cracked and spalled, and should be replaced.

the application of a carpet coat. Potholes can also become a nuisance and need repairing before they result in the disintegration of both the base and wearing courses.

VI Boundaries

6.1 Cast Iron Fence

The eastern boundary to the site consists of a tall privet hedge reinforced by a 6 ft (1.800) cast iron fence. The fence is in good condition apart from the rusting of the metal owing to lack of painting. The fence badly needs cleaning down, priming and painting for protection purposes.

The boundary walls and fences form substantial items and must be adequately covered in the report. The cast iron fence is corroded and the importance of preventing the formation of excessive rust and regular painting to preserve the fence cannot be over-emphasised.

6.2 Brick Walls

The other three boundaries to the site are bounded by one-brick walls, varying in height from $9'-0''$ to $12'-6''$ (2.74 to 3.81 m), in local bricks topped with bull-nosed coping bricks. These walls are the original ones surrounding the rear garden to 23 Sloane Street, and are barely of sufficient thickness, having regard to their length and height, and the absence of supporting piers. There is evidence that some repair work was undertaken on them when the depot was built in 1960.

One section of wall about $24'-0''$ (7.315) long and $10'-0''$ (3.048) high is bulging badly and is in a dangerous condition, probably caused by the four coniferous trees on the adjoining property. This section of wall needs rebuilding, and it would be advisable to erect three $9'' \times 9''$ (215 x 215) piers to further strengthen the wall, as it is retaining soil to a depth of about $2'-6''$ (750) on the opposite side.

The one-brick walls on three sides of the property must be a cause for some concern, having regard to their age and contruction. Walled gardens are very attractive until major repairs or even reconstruction of the walls becomes necessary.

The walls require careful examination in order to determine their structural condition and the extent of necessary repairs. Bulging, leaning or cracked walls all require careful scrutiny, to ascertain the likely cause(s) and the essential remedial work. It may be difficult to obtain matching bricks where the original locally manufactured bricks are no longer available. The walls may be jointed in lime mortar or

There are other areas of wall which are showing signs of deterioration. Essential repair work carried out now will avoid the need for much more costly repairs in the future. In all about 50 bricks are perished or laminated and these should be cut out and new matching bricks stitched in. In addition there are extensive areas of brickwork where the mortar joints are crumbling badly and these need raking out and repointing with gauged mortar (cement, lime and sand – 1:1:6), to prevent the ingress of moisture, frost action and subsequent damage to the brickwork. The total area involved is about 170 sq yds (142 m²).

About 80 ft (24.384) of coping bricks are loose and require cleaning and rebedding, before they become dislodged and dangerous, apart from ceasing to give the desired protection to the head of the wall.

the mortar may have perished, and extensive preparation and repointing, preferably with a gauged mortar, will be required.

Coping bricks or stones may be missing, loose or have cracked joints, which prevent them from performing their important protective role at the head of the wall effectively.

6.3 Gates

There are two sets of wrought iron gates, one at the start of the access road and the other at the entrance to the site. Both are in good working order.

VII Drainage

7.1 Foul Drainage

The discharge from the sanitary appliances in the mess room and lavatories flow directly into a manhole at the southern corner of the block. A foul drain runs from this manhole to a second manhole situated at the outlet end of a petrol interceptor at the western end of the paved area. The final length of foul drain connects the second manhole to a private manhole in the rear garden of 21 Sloane Street. The drainage is then conveyed through a 4″ (100) private drain laid

The first step in the survey of drains is to trace their lines. In this case, the manholes at each change of direction make this a simple task. The connection of the foul drains to a private house drainage system could be a matter for concern. The adequacy of the house drains to carry the increased flow and the question of liability for maintenance of

to an adequate fall and through a further manhole to connect with the public foul sewer in Sloane Street. A satisfactory drainage agreement has been obtained with the owner of 21 Sloane Street, who receives an annual payment of £30 for permitting drainage from the site through his drainage system but it does not include any liability for maintenance by him.

The drainpipes are 4″ (100) clay with flexible joints and are laid to self-cleansing gradients (about 1 in 50). The pipes were tested with a water test which was satisfactory. All the foul drains are very shallow and are surrounded with concrete.

The manholes are 3′-0″ × 2′-4½″ (914 × 702) internally and have been well constructed in one-brick walls in engineering bricks in cement mortar in English bond on a concrete base, and topped with a concrete slab and cast iron manhole cover. They are in good condition, apart from some cracked concrete benching at the base of one manhole and a cracked cast iron manhole cover −24″ × 18″ (600 × 450) − to the other.

the house drains are particularly important. Fortunately, the flow from the sanitary appliances is relatively small and the house drains are laid to a good fall. Furthermore, the discharges of vehicle washdown water from the petrol interceptor will help to flush the house drains.

It is necessary to establish the size and form of the drains and whether they are laid to adequate falls. The carrying out of a water test on the drains will show whether they are watertight.

The manholes need inspecting to determine their size, construction and condition. They can often be the cause of major blockages in the drainage system owing to loose rendering and other materials falling into the channels.

7.2 Petrol Interceptor

A three-chamber petrol interceptor has been provided to intercept oil and petrol arising from the washing of vehicles and to prevent their passage into the public sewers.

The petrol interceptor is constructed in similar materials to the manholes. The chambers are 3′-0″ × 3′-0″ (900 × 900) on plan and 3′-0″ (900) deep below water level. The outlet pipe from each chamber consists of a 4″ (100) diameter cast iron trapping bend. A 2″ (50) copper tube connects each chamber to a 4″ (100) cast iron ventilating pipe which is carried up above the roof level of the mess room and lavatories. This ventilating pipe also serves the head of the foul drainage system. The petrol interceptor is

The petrol interceptor is a very good feature and ensures the satisfactory disposal of any petrol or oil and the safe discharge of any dangerous fumes before they can cause explosions. The petrol interceptor is dealt with in a similar manner to the manholes, but including reference to the ventilating system. Any defects are duly recorded for inclusion in the summary of the cost of repairs.

in good condition and is operating effectively. The covers need rebedding in grease to prevent the egress of foul gases.

7.3 Surface Water Drainage

The surface water drains are extensive and consist of two main runs. One with three manholes passes down the south side of the site in front of the large garage block and collects the roof water from this block. The second leg starts with a manhole near the mess room and lavatories, passes in front of the small garage block and terminates near the eastern boundary of the site. This run also has three manholes and collects the roof water from the two buildings, and most of the paved area and access road within the site through four gullies. A drain crossing the eastern end of the site connects the two drain runs between the bottom two manholes.

The last length of surface water drain extends eastwards from the site for a length of about 96'-0" (29.26 m), across a gravel access way, allotments and a copse, and finally discharges into the River Lewis.

All the surface water drains are in clay pipes with flexible joints. All the main runs between the six manholes and the river are 6" (150) diameter and all the connections are 4" (100) diameter. The southern drain run is laid to a gradient of 1 in 30, the northern drain run to 1 in 60, the connecting length between the bottom manholes to 1 in 35 and the outfall to the river to 1 in 60. All the surface water drains are laid to self-cleansing velocities which prevent silting up, and they are of adequate capacity to cope with a heavy storm. The pipes were tested with a smoke test and showed no leaks, except for the length between the two bottom manholes, where there is a defective length at about the half-way point. It will be

The surface water drainage layout is extensive and it would probably help the client if a site plan were to be attached to the report showing the location of the manholes, gullies, lines of drains and the headwall to the river.

The same basic information about the drains is required as for the foul drains, including materials, sizes, gradients and lines of drains, and their condition. The manholes need only brief mention as they are of identical construction to those on the foul drains.

The water test used for the foul drains is considered excessively harsh for surface water drains and so a smoke test has been used.

The client will wish to know that the drains are laid to satisfactory gradients and are in consequence unlikely to give trouble in use through silting. Very few surface water drainage systems have sufficient capacity to cope with an exceptionally heavy storm of considerable duration, and in these unusual circumstances, some temporary flooding of the site could occur.

Any defects must be noted and the remedial work adequately described.

necessary to excavate and expose the drain and to rectify the defective length, to retest after carrying out the remedial work and then reinstate the excavation, which is in rough grassland.

The six manholes are of similar construction to those provided on the foul drainage system and are in good condition, apart from some cracked concrete benching to the bottoms of two manholes which need replacing, and two cracked manhole covers in areas crossed by vehicles. These two cast iron covers – 24″ × 18″ (600 × 450) – should be replaced with heavier duty covers.

The four road gullies are precast concrete trapped gullies, 15″ (375) diameter and 2′-6″ (750) deep, and all need cleaning of silt but are otherwise satisfactory.

7.4 Surface Water Outfall

The 6″ (150) surface water outfall discharges into the River Lewis through a 9″ (225) thick concrete headwall, supported by 9″ (225) wingwalls and a 6″ (150) apron. One of the wingwalls 3′-0″ × 4′-0″ (900 × 1200) is badly cracked and needs replacing. A cast iron flap valve is fixed over the pipe outlet to prevent river water backing up the drainpipe in times of flood. The valve is a very sturdy and effective one but needs lubricating for better operation.

A satisfactory pipe easement has been obtained for the outfall pipe to the river and involves an annual payment of £15.

The outfall to the river with the flap valve deserves mention. As the client is unlikely to be familiar with the form and purpose of flap valves, this information should be included in the report. Enquiries were made concerning the pipe easements for the surface water drain laid across private land to discharge into the River Lewis. Details are incorporated in the report in order that the position is made clear to the client and to ensure that he is aware of his financial obligations.

Surface water can be disposed of in a variety of ways according to the facilities available – public sewers, watercourses and soakaways. The surveyor will need to investigate the system fully, whatever method is employed.

VIII   Water Service

The water supply is provided by a ¾" (19) copper service pipe which was laid by the Newton Bassett Rural District Council through the garden of 21 Sloane Street from the water main in Sloane Street, A pipe easement agreement has been entered into with the owner of 21 Sloane Street and an annual payment of £15 is payable for this privilege. The service pipe enters the site in front of the mess room and lavatories and branches are taken to serve each of the three buildings with stopvalves on each of the connections.

Draw-off valves are provided in each of the garages for washing vehicles. All exposed pipes are in good condition, all valves are in satisfactory working order and the water is of adequate pressure. The stopvalve chambers give effective access to the valves and are painted yellow for ease of identification. The services pipes are laid at a depth of 2'-6" (750) which gives adequate protection from frost.

The water service enters the site through an adjoining property. Once again the surveyor had to check on the existence and terms of the pipe easement and to include the particulars in the report. The client should also be supplied with adequate information about the water supply arrangements on the site. Any defects or deficiencies in the supply arrangements must be included, so that the client is aware of any possible problems and limitations, and is aware of their consequences in practical and financial terms. Where the supply emanates from a private source, even more stringent investigations are required.

IX   Estimate of Cost of Repairs

The following is a summary of the repair works considered necessary to the property with their estimated costs.

Large garage   £

1. Clean down and paint 4 steel windows and steel angle heads and replace broken glass   160
2. Replace cracked roofing tiles to external window sills   15
3. Renew mastic pointing around windows   20
4. Clean down and paint all steelwork in stanchions, roof trusses and purlins   850

The surveyor should now systematically work his way through the draft report and prepare a schedule of all the items of repairs and replacements that are considered necessary, preparatory to pricing.

Brief but adequate indentifying descriptions are given of each component item. The estimated costs can be calculated in a variety of different ways or a combination of them.

(1) The recorded costs of similar previous items in the surveyor's office.

5. Replace six asbestos cement roofing sheets (note: if the whole of the roof sheeting and ridge capping were to be replaced, this would cost about £1500) ... 100
6. Clean out metal box gutter ... 10
7. Replace asbestos cement eaves gutter and four downpipes with aluminium rainwater goods ... 400
8. Replace timber doors with aluminium roller shutter doors, operated by endless hand chains ... 6000
9. Provide electric power and lighting installation ... 400

Small garage

10. Underpin 15'-0" (4.572 length of rear wall ... 300
11. Waterproof rear wall to a height of 3'-0" (900) ... 150
12. Repoint brickwork to verges ... 20
13. Clean down and paint all steelwork in stanchions, joists, channels, roof trusses and purlins ... 550
14. Replace all asbestos cement roof sheeting and ridge capping ... 750
15. Clean out metal box gutter ... 10
16. Refix and repoint lead flashing to rear gutter ... 15
17. Replace asbestos cement eaves gutter and four downpipes with aluminium rainwater goods ... 280
18. Replace timber doors with aluminium roller shutter doors, operated by endless hand chains ... 3200
19. Cut out and make good defective area of concrete floor ... 40
20. Provide electric power and lighting installation ... 350

Mess room and lavatories

21. Repoint bitumen-felt damp-proof course ... 15
22. Repoint defective brickwork to rear wall ... 50

(2) Calculation from first principles, taking the cost of the basic materials, cost of fixing and the appropriate additions for contractor's oncosts and profit.

(3) Extraction of appropriate prices from building price books, suitably adjusted to meet the site conditions, tendering climate and other relevant factors.

The items are best listed under suitable sub-headings, which act as signposts and help to lead the client through the various costs and assist in highlighting the areas generating the most costly repair and replacement work. The numbering of the items also helps for identification purposes.

Most of the cost estimates are rounded off to the nearest £10 as they cannot usually be that precise. When computing costs, it must always be remembered that small quantities of alteration and repair work are significantly more expensive per unit than larger quantities of new work in unrestricted locations. Hence considerable care is needed in arriving at the appropriate estimated costs in order that they shall be realistic.

Where areas or lengths are stated, both imperial and metric dimensions have been included, following previous practice, to help readers who may be conversant with only one or the other. At some time in the future it is to be hoped that we

23. Repair lengths of defective asphalt skirting to roof	30
24. Clean down, prime and paint metal windows	100
25. Replace two internal doors	80
26. Replace three sets of lever door handles	30
27. Cut out and replace defective plaster	40
28. Clean down and repaint all internal painted surfaces	700
29. Replace cracked WC pan	45
30. Replace defective ball valve	5
31. Replace two badly crazed wash basins	120
32. Replace 8 lighting points and switches and 6 power socket outlets	70

Paving and access road

33. Clean out expansion joints to access road and paved areas and refill with sealing compound	180
34. Replace three bays of defective reinforced concrete access road	1680
35. Make good damage to kerbs after relaying concrete	20

Boundaries

36 Clean down, prime and paint cast iron fence	900
37. Take down and rebuild 24 ft (7.315) length of one-brick boundary wall	1360
38. Construct three 9″ x 9″ (215 x 215) brick piers to boundary wall	150
39. Cut out and stitch in 50 bricks	60
40. Repoint 170 m² of brick facework	1350
41. Clean and rebed 80 ft (24.384) of coping bricks	70

Drainage

42. Repair cracked benching to one foul manhole	5

shall all be using the same system of measurement, and have moved away from the present unsatisfactory situation.

It will be noted that the schedule includes items of additional work, such as the electric power and lighting installations to the two garage blocks, where these are considered necessary for the client's intended purpose.

43. Replace one cracked manhole
    cover to foul drains   35
44. Rebed three covers to petrol
    interceptor in grease   10
45. Excavate to expose, replace
    defective length of 6″ (150)
    surface water drain and reinstate   200
46. Repair cracked benching to two
    surface water manholes   10
47. Replace two cracked surface water
    manhole covers with heavy duty
    covers   105
48. Clean out four road gullies   10
49. Replace wingwall to surface
    water outfall to river and
    lubricate flap valve   50

Total estimated cost   £21 100

X Conclusions

The site has considerable potential for further development, with about one-half of the plot of ½ acre (0.2 ha), which is adequately serviced and accessible but not built upon. Enquiries at the local authority offices did not indicate that there would be any major difficulty in obtaining the necessary building regulation and planning approvals for the erection of additional buildings for the maintenance and garaging of vehicles.

The site occupies a central location in the town and should not give rise to any major security problems. The approach roads to the site are rather tortuous and the first section of communal access road could result in maintenance problems because of the joint responsibility with other property owners.

The connection of the foul drain into a private house drainage system, and the laying of the water service and surface water outfall across private land, are all features

The report normally concludes with the surveyor's conclusions which will give his opinion on the general condition and suitability of the property for the client's intended use. He may also be required to give a valuation of the property and an estimate of the cost of essential repairs and replacements. Valuations are always difficult in the absence of recent comparables, particularly for one-off type buildings such as those under consideration.

The potential of the site for further development should be explored so that the client is made aware of the possibilities for future extensions to the present buildings. This could be an important aspect.

which are best avoided where possible, but are not considered matters for concern. However when the site was developed in 1960 there were no feasible alternatives and all necessary safeguards have been taken. They do however result in a total additional annual outgoing of £60.

In general the buildings are well constructed and meet your stated needs very well. The major deficiencies in the garage blocks are identified in the report and relate mainly to the condition of the roofs, doors and paintwork and the absence of electrical power and lighting. There are also defects in the brickwork to the small garage.

The repairs to the mess room and lavatories cover a variety of items, but once these are undertaken and some suitable furniture installed, the building will serve its purpose well. Further improvements would be the provision of showers and a refrigerator, and possibly a small office.

Some significant repairs are needed to the boundary walls and the concrete access road. The services generally are in good, serviceable condition.

If you decide to purchase the property, it is strongly recommended that the listed repair and replacement work be undertaken to put the property into a satisfactory condition and to prevent further deterioration. I shall be pleased to discuss any of the suggested works with you and to consider possible alternatives if required.

I value the property in good condition at £152 000, and this includes the land and services. To arrive at its current market value in its present condition, it is necessary to deduct the cost of the essential repairs at about £21 000, giving a net figure of £131 000. The asking price is £145 000 and I would recommend that an offer of £130 000

Any major drawbacks connected with the property should be stated, in order that the client shall be fully aware of all the disadvantages and how they are likely to affect his occupation of the property.

The repairs, although extensive, are not so serious as to make the property unacceptable. They can all be rectified but at a cost, in excess of £20 000. The cost of the repairs must be taken into account when deciding the sum that can be offered for the purchase of the property.

Conclusions often contain a brief summary of the main components of the report. There can however be no one set approach and much will depend on the type of property and the approach adopted by the individual surveyor. It forms very much a personal document from a surveyor or other professional to his client and its format will be influenced by the client's needs and the terms of reference which he gives to the surveyor. There is, therefore, no such thing as a typical report. Finally, the report should conclude with the surveyor's name or signature and the date when the report was prepared.

be made in the first instance if you decide to
proceed with the purchase of the property.

J. P. Charterhouse FRICS
Chartered Surveyor

27 March 1985

Report on the Structural Condition of a Dwelling House

The following report on a pre-war house is kept concise and relatively free of
technical terms and concentrates on the main issues. It is reasonably self-explana-
tory and sidenotes have accordingly been omitted. The report is preceded by a
covering letter. The approach varies from the example given in chapter 4 to show
an alternative method, as there is no one universal system.

Johnson and Wagstaff
Chartered Surveyors
54 High Street
Greater Horton
Buckinghamshire
20 March 1985

John Hustlewaite Esq.
'Handel'
Marsh Road
Congleworth
Norfolk

Dear Sir

'Gladwyns', 38 Blueberry Road, Lesser Horton

Further to our letter of 5 March, detailing our terms of reference, we have
pleasure in enclosing our report on the structural condition of the above property. We shall be pleased to amplify any matters in the report should this be
required.

Yours faithfully
Johnson and Wagstaff
Enclosure

'GLADWYNS', 38 BLUEBERRY ROAD, LESSER HORTON
REPORT ON STRUCTURAL CONDITION OF PROPERTY

1. General

The house is detached with two storeys and is situated on a plot with a frontage of 48 ft (14.8 m) and a depth of 125 ft (38.5 m), in a good class residential area. The front stone boundary wall is the liability of the owner and is in satisfactory condition. The other three boundaries are of wood close boarded fencing in fair condition. The owner is liable for the fence adjoining No. 40. The garden is well served with concrete paths but the garden itself has been neglected.

Blueberry Road is in good condition and is maintained by the Swindle-hurst District Council. Gas, water and electricity services and a combined sewer are located in the road.

2. Outbuildings

A timber framed and asbestos sheeted single garage is situated at the rear of the house, with a satisfactory approach to the road. Some of the asbestos sheeting to the garage is cracked and requires replacing. A small timber summer house at the end of the back garden is in satisfactory condition but requires an application of wood preservative.

3. Main Structure

The house was built in 1928 and has an effective damp-proof course. It is constructed of brick cavity walls, 11"(255) thick, faced with multi-coloured facing bricks, except for the south elevation which is partly rendered. The brickwork to the north and west elevations requires pointing as the mortar joints are crumbling. The cement rendering to the south elevation is badly cracked and loose in parts and requires stripping and renewing.

The roof is covered with clay tiles, some of which have laminated and require replacing, as there is evidence of rain penetrating the roof space. There is 2" (50) of insulation over the first floor ceiling. The roof timbers are sound and of adequate dimensions. Two cracked lengths of cast iron gutter and one defective length of downpipe need replacing.

There are two chimney stacks, one of which is cracked, unsafe and requires rebuilding. The other stack is sound but needs repointing, and the cement flaunching around the chimney pots renewing.

Mastic pointing around the door and window frames is needed to keep them watertight. Defective putties to six windows need replacing. The paint-work to all external woodwork and metalwork is in very poor condition and urgently needs repainting, including thorough preparation.

The floors are constructed of softwood boarding on timber joists of adequate dimensions. All floors with the exception of bedroom 2 are in sound condition and the ground floor is adequately ventilated with air bricks. The floor to bedroom 2 has two defective joists which require replacing.

4. *Internally*

Ground floor

Entrance hall, lounge and dining room. The wood doors and windows, and plastered walls and ceiling are all sound but the decorations are in only fair condition.

The fireplaces in the lounge and dining room are of satisfactory construction, but the one in the dining room smokes badly and appears to be blocked with debris. This has fallen from the defective chimney stack and can be remedied when the stack is rebuilt.

Kitchen and larder. The wooden doors and windows, partly tiled and plastered walls and plastered ceilings are all in sound condition but the decorations are poor. The kitchen is fitted with a gas cooker which although of an old type is in satisfactory working order. The ceramic sink is chipped and the teak draining board is rotting in two places. Replacement with a modern sink unit is desirable. There are no fitted cupboards, but adequate storage space exists in both the kitchen and larder.

First floor

Staircase and landing. The staircase and landing are structurally sound apart from three cracked balusters which need replacing. Access to the roof space is by a hinged trap door of adequate dimensions in the landing ceiling. The roof space is partly boarded and provides considerable storage space.

Bedrooms 1, 2 and 3. The three bedrooms are structurally sound apart from the floor to bedroom 2 (see 'Main Structure'). All decorations are in reasonable condition and no work is necessary unless it is desired to change the colour schemes. Some making good of finishes will be needed after the floor repairs are carried out in bedroom 2.

Bathroom and WC. These are both fully tiled and no decorations are necessary apart from the plastered ceilings, where cracks in the plaster need cutting out and making good.

The bath is cast iron, enamelled and sound and needs replacing only if a more modern fitting is required. The wash basin is of an old design and slightly chipped; replacement is optional. The WC is a modern low level coloured suite in good working order.

5. *Plumbing*

The cold water rising main is taken in a lead service from the Water Authority's main to a storage cistern in the roof space. The water services are in good condition but the cistern is badly corroded and needs replacing, before it starts to leak.

The domestic hot water is supplied from an electric immersion heater in the main cylinder and operates efficiently. The pipework between the back boiler in the dining room and the main cylinder was removed by a previous owner.

6. *Heating*

The lounge, dining room and bedrooms 1 and 2 have fireplaces. In addition there are gas points to the lounge and dining room fireplaces, which are capped off, but gas fires could be installed if required.

7. *Electrical Installation*

The house is wired throughout in metal conduit to serve electric lights and power socket outlets, which are provided to an acceptable level. There are satisfactory fuse boxes and cut outs and the electrical installation has been tested by the Electricity Board and found to be in order.

8. *Drainage*

The drainage system was water tested and found to be satisfactory. Manholes and gullies are silted up and need cleaning. One cracked manhole cover requires replacing.

9. *Estimated Cost of Repairs*

The following itemised list shows the estimated cost of putting the property into a satisfactory state of repair.

		£
(1)	Replacing cracked asbestos cement sheets to garage	110
(2)	Application of wood preservative to summer house	20
(3)	Repointing brickwork externally	700
(4)	Stripping and renewing cement rendering to south elevation	230
(5)	Replacing defective roofing tiles	45
(6)	Renewing eaves guttering and downpipe	65
(7)	Taking down defective chimney stack and rebuilding	260
(8)	Repointing chimney stack and renewing flaunching	40
(9)	Replacing two defective floor joists and making good associated work	60
(10)	Replacing sink and drainer with new sink unit	115
(11)	Replacing three defective balusters to staircase	20

(12)	Replacing cold water storage cistern in roof space	50
(13)	Cleaning out manholes and gullies	10
(14)	Replacing cracked manhole cover	70
(15)	Renewing mastic pointing around door and window frames	45
(16)	Reputtying glass to six windows	40
(17)	Internal decorations	200
(18)	External painting	230
	Total	£2310

10. Conclusions

From our inspection we formed the opinion that this house could be a sound and comfortable dwelling, although it is lacking some of the usual amenities, such as central heating, modern sanitary appliances, kitchen fitments and built-in bedroom fittings. These items can be added subsequently. The carrying out of the listed repairs will make the house sound and habitable.

signed . . . .
date . . . .

Alterations and Improvements

Surveyors are frequently requested to advise on the modernisation, adaptation and alteration of properties. The first step is a survey of the property on the lines described in chapter 4, when its potential for adaptation will be fully considered, having regard to its stability, suitability for conversion and all relevant statutory controls, including planning permission, open space, ventilation, protection of drains and fire regulations. Drawings will then be prepared showing the proposals, accompanied by a report. Projects may vary from the modernisation of pre-war dwellings to the conversion of schools into workshops and churches into car showrooms. A cost study may be necessary to ensure that the cost of acquisition and conversion does not exceed the cost of land purchase and construction of a new purpose-built structure.

Older dwellings need careful consideration and skilful planning to make the fullest use of their potential for improvement and conversion without at the same time adversely affecting their character. For example, sound doors and windows should not be replaced by inappropriate modern fittings, nor should pleasant large rooms be divided unnecessarily, nor small sunny gardens be drastically reduced by kitchen and bathroom extensions.[7]

Wider frontage dwellings provide scope for inserting a bathroom without losing bedspace, incorporating an entrance hall, or providing an extension without

overshadowing rear rooms or garden. Deeper dwellings often permit the economical provision of internal bathrooms. A second entrance to a dwelling may be
closed to provide more usable floor space on the ground floor. Stairs may be
reconstructed to provide increased safety or a more effective first floor layout.[7]

After improvement, dwellings should normally possess a minimum life of
30 years and careful thought and attention should be given to the following
basic criteria.

(1) A main entrance door opening into an entrance porch or hall.
(2) Conveniently arranged bathroom, internal WC and spacious kitchen.
(3) The provision of independent access to all rooms from circulation areas,
     except possibly to a kitchen from a dining or living room.
(4) Ideally, the provision of two living spaces in family dwellings, although in
     small houses one of these could be a reasonably spacious dining kitchen.
(5) Use of appropriate materials and components externally to ensure harmony
     with existing elevations.[7]

A survey of modernisation schemes for pre-1945 houses in 17 former county
boroughs in England and Wales showed the following operations to be the most
common:

(1)   plastering of fair-faced brickwork;
(2)   renewal of floors and laying of thermoplastic/PVC tiles;
(3)   removal of old fireplaces from living room and bedrooms;
(4)   installation of central heating systems;
(5)   renewal of electric wiring;
(6)   installation of 13 amp power points and 30 amp cooker control unit;
(7)   provision of new sink unit:
(8)   removal of larder and provision of kitchen floor units and ventilated wall
      cupboards;
(9)   provision of new bath, wash basin and WC suite; and
(10) removal of small paned steel windows.[7]

References

1. A. E. Darbyshire. *Report Writing.* Arnold (1970)
2. Royal Institution of Chartered Surveyors. *Guidance Note on Structural
   Surveys of Commercial and Industrial Property* (1983)
3. B. M. Cooper. *Writing Technical Reports.* Penguin (1964)
4. Royal Institution of Chartered Surveyors. *Structural Surveys of Residential
   Property: A Practice Note* (1981)
5. D. Chappell. *Report Writing for Architects.* Architectural Press (1984)
6. M. Gass. How to write better survey reports. *Chartered Surveyor* (October
   1981)
7. I. H. Seeley. *Building Maintenance.* Macmillan (1976)

6 Comparison of Structural and Technical Reports, Schedules of Condition and Proofs of Evidence

This chapter makes a comparative study with supporting examples of the different types of report prepared by the surveyor, their purpose and the way in which they are used.

Structural Reports

Two examples of structural reports were included in chapter 5, covering two different building types and illustrating variations in approach. Earlier in the book, it was shown how the terms 'structural survey' and 'building survey' are used in practice often without any clear distinction between them. Essentially a structural report provides a detailed account of the condition of a building with a full description of the defects and their significance. The report often contains an estimate of the cost of carrying out essential repairs and an opinion on the suitability of the building for the client's intended use. On occasions it is also extended to cover possible alterations and adaptations where these are required by the client.

For example, a director of housing to a local authority may be requested to report on the structural, sanitary and environmental condition of poor class pre-war houses, which the local authority may be considering purchasing and bringing up to a certain minimum standard. In these cases it is necessary to carry out a very thorough investigation as the number and scope of defects can be extensive. The following schedule of typical defects likely to be encountered in this class of property indicates the nature and scale of the investigation.

External

Walls: dilapidated, bulging or dangerous walls
        defective or perished bricks or masonry
        defective or perished pointing
        defective rendering
        defective or eroded sills
        defective arches
        defective or missing pointing around door and window frames
Damp-proof courses: absent or defective damp-proof course
                     soil or paving above damp-proof course
Ventilation to suspended
ground floor: insufficient or non-existent ventilation
        defective or blocked air bricks
Roofs: dilapidated or inadequate roof timbers and/or
        roof coverings to pitched and flat roofs
        defective, missing or loose slates or tiles
        defective flashings or cement fillets
        defective ridge, hip or valley coverings or parapet walls
Chimney stacks: dangerous and/or defective chimney stacks
              defective brickwork or pointing
              broken or missing chimney pots
              perished or cracked flaunching to chimney pots
Gutters: absence of adequate gutters
        defective or choked gutters
Rainwater downpipes: absence of rainwater downpipes
                  defective or choked pipes
                  defective rainwater heads or shoes
                  short lengths of pipe discharging on to walls or other surfaces
Drainage: defective or inadequate drainage system
        insufficient and/or defective manholes
        absence of proper ventilation
        defective gullies
        defective soil pipes, vent pipes and associated fittings
Paving: unpaved surfaces to yards and other vital areas
        defective paving
        paving above damp-proof course or
        ground floor causing dampness
Boundaries: dilapidated or missing fences
            defective boundary walls

Internal

Walls and ceilings: defective, damp or perished wall plaster
                  defective wall pointing to fair-faced work

defective, bulging or damp ceilings

dirty walls and ceilings and/or poor decorations

Windows; defective or decayed frames, sashes, beads, sills and other associated members

lack of opening lights

insufficient window area

broken or missing sash cords

defective fastenings

cracked or loose glass

Doors: defective or decayed woodwork to doors and/or frames

absence of or defective weatherboards

defective thresholds

defective door furniture

Floors: dilapidated, defective or loose floorboards or joists

defective or damp solid floors

Stairs: inadequate and/or insecure stairs

defective treads, risers, nosings, handrails, balusters and other members

Fireplaces: defective kitchen ranges, firegrates, firebacks, flues, hearths and other components

Larders: absence of adequate ventilated food stores

Sanitary appliances: absence of separate or suitable WC accommodation

unsatisfactory or dilapidated WC compartments

defective WC pans

defective or insanitary woodwork around pans

defective flushing cisterns, seats, chains and other apparatus

defective sinks, baths or wash basins

defective traps or waste pipes

absence of suitable impervious surrounds to appliances

Heating: inadequate heating system

lack of or inadequate thermal insulation

Water supply: inadequate or unsuitable water supply

inadequate or defective pipework and fittings

Electricity and
gas supplies: unsatisfactory or inadequate wiring or pipes

unsatisfactory or inadequate fittings.

Technical Reports

Building Defects

Apart from the more usual type of structural report, a surveyor may be requested to prepare a report on a specific problem, such as a building defect. The following example of a typical report of this nature will help the reader to appreciate its possible scope and format.

Background Information

A substantially brick-built house has a parapet wall supporting a lead parapet gutter to a tiled roof. The owner of the property, who is responsible for repairs, has received a complaint from the tenant that rain has caused a damp ceiling in the front bedroom. The owner has commissioned a surveyor to report on the matter. In this case the report is incorporated in a letter to the client.

Report

F. B. Appelby Esq.   25 April 1985
The Old Vicarage
Lesser Hambury
Suffolk

Dear Sir

Dampness at 38 Elm Walk, Beestonfield

Cause of complaint
Following the complaint of dampness in the front bedroom of the above house, I examined the parapet wall, gutter and adjoining roof on 24 April 1985 and found them to be soundly constructed. Several defects have however developed which have resulted in rainwater penetrating the front bedroom ceiling. These defects are as follows.

(1) A length of lead flashing forming the junction between the parapet wall and the lead gutter has come away from the wall, thus permitting rainwater to pass down the inner face of the wall at this point.
(2) A length of lead gutter has been damaged by some sharp object penetrating the lead, possibly the feet of a ladder.
(3) The gutter has become badly choked with decaying leaves and other debris, causing water to build up in the gutter and ultimately flow over the top edge of the lead covering to the gutter, where it passes up the tiled roof slope. The problem has been aggravated because the lead gutter covering does not extend as far up the roof slope as is really desirable, nor is it dressed over a tilting fillet (triangular piece of timber) below the bottom course of roof tiling, which would have produced a much sounder job.

Remedial work
I recommend that the following remedial work be carried out by a selected builder.

(1) Hack out the brick joints and rewedge and repoint the loose length of lead flashing.
(2) Replace the defective length of lead gutter covering.
(3) Clear all debris from the gutter.

It is not considered necessary to replace the lead gutter covering with wider sheets or to fit a timber tilting fillet, provided that the gutter is cleaned out regularly, preferably annually.

The estimated cost of carrying out the repair work to the parapet gutter is £95. Additionally, the decorations to the ceiling of the front bedroom are badly discoloured and need redecorating with two coats of emulsion paint, after the external repairs have been carried out and the plasterboard ceiling has adequately dried out. The estimated cost of the redecorations is £35.

R. S. Harrington
Chartered Surveyor

Report on Drainage and Sanitary Arrangements

Figure 6 illustrates the drainage system to a house with a number of defects to the installation and the sanitary appliances listed on the drawing. The house, situated on the outskirts of a small country town, was built in about 1930 and contains four bedrooms. The subsoil is clay.

The client requires a report detailing any remedial works considered necessary, together with an approximate estimate of their cost.

Report on Drainage System and Sanitary Arrangements at 'Gladways', Hamlet Road, Toftmead, Norfolk, for Col. P. T. Featherstone-Jones
1. *Drainage System*
   The discharges from all sanitary appliances are drained into a septic tank and filter, for purification purposes, and then flow into the ditch adjoining the property on the north side, through a small headwall. Rainwater from the front of the house discharges into the ground and that from the rear drains into a rainwater tank.
2. *Drainpipes*
   All drains consist of 4″ (100) clay pipes with cement mortar joints laid to satisfactory gradients. The length of drain (30 ft − 9.3 m) between the second manhole and the ditch is leaking badly. It is necessary to excavate to expose the drain, and it will probably be advisable to replace the full length of drain with new clay pipes with flexible joints. The pipes should then be tested with a water test before the trench is backfilled. This length of drain passes through a vegetable plot and a small copse. It is possible that tree roots have entered the pipe joints in search of moisture. The drains

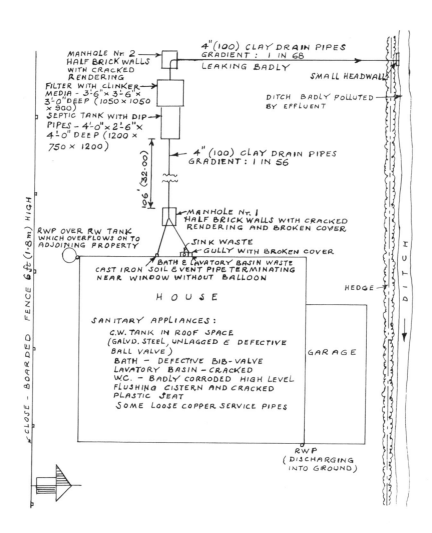

Figure 6. Drainage plan of house

from the house to the septic tank installation withstood a water test satisfactorily and are in sound condition.

3. *Manholes*

The two manholes located at drain junctions and changes in direction are constructed of 4½" (102) brick walls in common bricks in cement mortar, rendered internally with a coating of cement and sand. The rendering to both manholes is cracked and requires hacking off and replacing before aggressive drain air attacks the brick walls, which being constructed of common bricks are particularly vulnerable. The cast iron cover to the first manhole is broken and needs replacing for reasons of health and safety.

4. *Septic Tank Installation*

The effluent from the installation is causing severe pollution of the ditch into which it discharges. The flow in the ditch is small except in times of moderate to heavy rainfall and so the discharges normally receive only a small amount of dilution by surface water. The bed of the ditch is badly discoloured and there is an offensive smell which is objectionable to occupants of the house and neighbouring properties in warm weather and could create a health hazard.

It is always difficult to secure a good quality effluent from an individual house septic tank installation. However, an examination of the existing installation showed the following major deficiencies.

(1) The septic tank is half filled with sludge (heavy solids), thus largely negating the action of decomposition by anaerobic bacteria. It has not been cleaned out for a long time and should ideally be emptied of sludge every six months.

(2) The clinker in the biological filter has largely disintegrated into dust, resulting in little oxidation of the septic tank effluent by aerobic bacteria. The two way tipper distributor in the filter is corroded and jammed, resulting in all discharges taking place over the surface of one-half of the filter.

(3) The worst aspect is, however, the inadequate size and poor condition of both the septic tank and the filter.

The septic tank is 4'-0" x 2'-6" x 4'-0" deep (1.200 x 750 x 1.200); whereas current Building Regulations require a minimum volume of 2.7 m³, the existing tank has a capacity of only 1.08 m³. Furthermore, it is a 4-bedroom house and could accommodate up to six persons. It would therefore be prudent to have a tank size in excess of the minimum, and we recommend dismantling the existing tank and replacing it with a large one — 10'-0" x 3'-4" x 4'-0" deep (3.000 x 1.000 x 1.200) giving a volume of 3.6 m³. The tank should be constructed with a base sloping gently towards the inlet dip pipe and be emptied of sludge twice a year for effective operation. The tank should desirably be constructed in 9" (215) brick walls in class B engineering bricks and be covered with removable precast concrete slabs.

The biological filter is $3'-6'' \times 3'-6'' \times 3'-0''$ deep ($1.050 \times 1.050 \times$ 900) giving a capacity of 1 m^3. A desirable capacity can be based on 1 m^3 per person for small works, indicating the gross inadequacy of the existing filter. A more realistic size would be $12'-0'' \times 6'-0'' \times 3'-0''$ deep ($3.600 \times 1.800 \times 900$), giving a capacity of 5.83 m^3. The new filter should be filled with hard-crushed gravel for durability, and fitted with a good quality tipping trough, feeding a series of fixed channels of metal or precast concrete with notched sides, and be adequately vented. The filter should also be constructed of 9" (215) brick walls in class B engineering bricks for durability.[1]

5. *Gully*

The gully taking the sink waste has a broken cover which should be replaced before it disintegrates, leaving a dangerous hole and possibly blocking the drain, and the gully requires cleaning of silt.

6. *Soil and Vent Pipe*

The cast iron soil and vent pipe terminates near a window and, apart from non-compliance with the current Building Regulations, could give rise to conditions which are prejudicial to health. The pipe requires extending upwards by a length of 4 ft (1.200), finishing above eaves level. The top of the pipe should be protected by a strong wire cage to prevent birds nesting in the pipe and so rendering it ineffective for ventilation purposes.

7. *Rainwater Disposal*

The rainwater from the rear of the house roof discharges through a rainwater downpipe into a cast iron rainwater tank with a boarded cover. When full the tank overflows into the garden of the adjoining property. This constitutes a tort (nuisance) in respect of which the owner is liable for any damage which the escaping water may cause. One remedy is for the overflow to discharge into a soakaway, $4'-0'' \times 4'-0'' \times 4'-0''$ deep ($1.200 \times 1.200 \times 1.200$), filled with suitable hardcore and covered with a precast concrete slab. We believe that a better alternative would be to relay the eaves gutter to fall to the north end of the house, to relocate the rainwater downpipe and rainwater tank at that end of the house and to connect the overflow from the rainwater tank to the ditch with a 12 ft (3.600) length of drain. This solution may sound more complicated but there must be doubt as to the efficiency of a soakaway in a clay subsoil.

The rainwater from the front roof slope discharges through a rainwater downpipe into the ground, where it has resulted in a waterlogged area of gravel drive and flower bed. If allowed to persist this could result in damp penetration into the house and damage to the foundations, apart from the inconvenience being caused. The most economical solution would be to connect the downpipe to a back inlet gully and to lay a 12 ft (3.600) length of 4" (100) clay drainpipes with flexible joints from the gully under the gravel drive to the ditch.

8. *Sanitary Appliances*

The sanitary appliances and service pipes were examined as requested and the solid fuel boiler in the kitchen, the hot water cylinder, cold water storage tank, electric immersion heater and copper service pipes are all of adequate capacity and operating effectively. A few loose lengths of pipe require additional supports.

There are several deficiencies in the sanitary appliances which require attention and these are now listed:

(1) The galvanised steel cold water storage tank in the roof space is unlagged and therefore vulnerable to damage from frost and the ball valve is defective. We recommend that the sides and top of the tank be lagged with polystyrene insulation and that the ball valve be replaced.

(2) There is a defective bib-valve (tap) to the bath which requires replacement. It might be considered advisable to replace both hot and cold water taps to the bath with matching modern fittings.

(3) The wash basin in the bathroom is of old pattern, supported by cantilever brackets and is cracked. We recommend that this appliance be replaced by a modern pedestal wash basin.

(4) The WC is located in a separate compartment. It is an old fashioned high level white appliance. The cast iron flushing cistern is badly corroded and the plastic seat to the WC pan is cracked. We recommend that this appliance be replaced with a modern coloured low level WC suite, which apart from being a much superior and more attractive fitting will assist in enhancing the value of the property.

9. *Estimate of Cost of Repairs*   £

(1) Replace drainpipes from second manhole to ditch	360
(2) Repairs to two manholes	160
(3) Reconstruct septic tank and filter in accordance with the recommendations contained in the report	3300
(4) Replace broken gully cover and clean out gully	5
(5) Extend soil and vent pipe and fix wire cage	30
(6) Reconstruction of rainwater disposal arrangements with two outfalls into the ditch	360
(7) Provide additional supports to copper service pipes	10
(8) Lag top and sides of cold water storage tank with polystyrene insulation and fit new ball valve	20
(9) Fit two new bib-valves to bath	15
(10) Replace wash basin with new pedestal appliance	90
(11) Replace WC with modern low level suite	160
Total	£4510

10. *Conclusions*

The major defect in the drainage system relates to the inadequacy and poor construction and maintenance of the septic tank and filter. The present unacceptably low standard of the effluent (discharge) from the installation is causing serious contamination of the ditch which passes through a residential area. This is a very serious matter from a public health viewpoint and requires urgent rectification. Unfortunately, the cost of providing a replacement installation of adequate capacity is an expensive undertaking.

Enquiries at the local authority's offices indicate that no public sewerage system is likely to be provided in this area for at least ten years. The only other alternative, which is not favoured by the local authority and whose approval is required, is the construction of a large cesspool for storage of the soil and waste discharges from the house and its emptying at frequent intervals. We do not believe this to be a satisfactory alternative but are prepared to discuss it with you should you so desire.

Other considerable items of expenditure relate to the new length of drain from the second manhole to the ditch, manhole repairs and the revised rainwater disposal arrangements, all of which are vitally necessary. The replacement of the WC and wash basin with modern appliances will increase the amenities and attractiveness of the house.

Hence the major part of the expenditure advocated in this report is required to carry out essential major repairs to the drainage system. It is evident that the heavy contamination of the ditch adjoining the north side of the property and the uncontrolled discharge of rainwater from the roof of the house cannot be allowed to continue without resulting in serious consequences. We accordingly strongly recommend that the remedial works listed in this report be undertaken as a matter of urgency.

J. F. Harrison and Partners
Chartered Surveyors
9 May 1985

Schedules of Condition

A schedule of condition is probably best described as a report on the condition of a property at a specified date, set out in sufficient detail so that any part of the structure, finishings or fittings which subsequently become defective or missing can be readily identified.

It is a normal condition of a lease or tenancy agreement that the lessee shall return the property to the freeholder on the expiration of the lease in a comparable condition to that appertaining at the commencement of the lease. However, in the case of a long lease, such as for 21 or 50 years, a considerable amount of wear and tear will have occurred for which adequate allowance must be made. A commonly adopted clause in such a lease reads 'fair wear and tear excepted', and then the main problem stems from the identification of what is fair in the particular circumstances, having regard to the condition of the property and the type and length of use. The lessee may also have carried out certain alterations and improvements to the property, but he may, nevertheless, be required to restore the property to the freeholder in similar form to that obtaining when he took the lease of the property.

It is a useful safeguard to both parties to the lease to have a professionally prepared and jointly agreed schedule of condition of the property at the commencement of the lease to prevent subsequent disputes. In the case of commercial premises a lessee could be faced with a substantial claim for diminution in value at the expiration of the lease. A lessee can be protected at the commencement of the lease if it is stated that he will not have to hand the demised premises back to the lessor on the expiration of the lease in any better condition than it was at the commencement, as evidenced by a schedule of condition. In these circumstances an agreed schedule of condition is prepared by the lessee's and lessor's surveyors, signed by both parties and attached to both the lease and the counterpart.[2]

When a surveyor takes instructions to inspect a leasehold property, he should enquire as to the existence of a schedule of condition, and if available he should desirably have a copy with him when he makes his inspection, so that he can compare his findings with the conditions described in the schedule.

Where a client proposes to take a lease of an old building in poor repair, the surveyor should recommend that a schedule of condition be prepared and agreed with the lessor prior to the exchange of the lease and the counterpart. This is especially important where it is a full repairing lease and the building shows obvious signs of structural weakness, settlement and other serious defects which can prove costly to rectify. Admittedly, the lessee's obligation is limited to repairs and does not extend to renewing or rebuilding the fabric. In practice the distinction between the two may not always be clear cut.[2]

A roof covering may be in such poor condition that it should be stripped and renewed to give satisfactory performance. In the short term it could perhaps be argued that limited patching could suffice and this could then form a contentious issue between the lessor and lessee. A schedule of condition detailing the state of the roof covering at the commencement of the lease will assist in providing a standard against which the reasonable extent of the remedial work can be judged.

Older buildings often suffer from dampness and the schedule of condition should identify any dampness occurring at the commencement of the lease and the existence or otherwise of effective damp-proof courses. These particulars will help considerably in determining the lessee's liability for subsequent damp penetration. Similarly cracked, bulging and leaning walls all require careful description in a schedule of condition to avoid the lessee being faced with excessive claims for rebuilding walls or underpinning at the termination of the lease. These examples serve to illustrate the value of schedules of condition in helping to determine the respective liabilities of both parties at the expiration of the lease.

Format of Schedules of Condition

A schedule of condition of domestic property is often best set out in the form of an abridged specification, possibly subdivided into three columns. The first column contains the categories of work in the form of sub-headings, the second column accommodating the descriptions of the particular items and the third column being used for their condition. Descriptions should be prepared in considerable detail and table 6.1 illustrates the use of this approach to an upper floor bedroom.

Table 6.1 Tabulation of schedule of condition to upper floor bedroom

Work classification	Description	Condition
Floor	Unpainted softwood boarding with stained margin	Three shrinkage cracks and one loose floorboard
Ceiling	Plastered, lined and emulsion painted	Four cracks in plaster and one damp patch of about 2 sq ft (0.2 m^2)
Walls	Plastered and papered	Large dirty patch on north wall
Windows	1¾″ (45) softwood casements with softwood frame and BMA fittings, glazed with clear sheet glass	One cracked pane of glass
Door	1½″ (38) 4 panel moulded both sides softwood door with mortice lock and BMA lever handles	Defective door furniture

A schedule of condition prepared in this way will clearly establish the condition of the property at the commencement of the lease, and is of considerable assistance to a surveyor preparing a schedule of dilapidations, of the form illustrated in chapter 8, at the end of the lease, when he can prepare a schedule of condition of identical format and compare the contents of the two.

An example follows of a schedule of condition of a commercial property, where it is not considered necessary to describe the construction, finishings and fittings in quite the same detail as was adopted for the upper floor bedroom.

Schedule of Condition of Premises on the north side of Coxham Road, Newtown, owned by J. S. Blackall; prepared on 17 June 1985
General Description
The property consists of three warehouse buildings surrounding a paved yard. The building to the east of the yard: Nr 1 (ground and first floors) is constructed

of brickwork part cement rendered, and a pantiled roof; that to the west of the yard: Nr 2 (ground and first floors) is constructed of brickwork cement rendered, and a boarded and felted roof; that to the north of the yard: Nr 3 (first floor only) is constructed of brickwork cement rendered, and a pantiled roof.

External Condition

General Condition

Poor

Cement Rendering

Fairly poor condition. Several cracks and slightly uneven in parts. Two small loose areas.

Brickwork

The south gable wall and west wall of building Nr 1 require repointing and 14 bricks have perished.

Woodwork

Poor condition. All the doors and sills to six windows are partly decayed. All woodwork is deteriorating rapidly because of lack of paint. Two doors are badly cracked. Four casement windows are missing and 48 small panes of glass are broken or missing.

Paintwork

Very poor.

Felted Roof

Requires a coat of bitumen and the skylights are leaking and require reputtying.

Tiled Roofs

Two buildings (Nrs 1 and 3) are leaking in places, because of the poor condition of the battens and torching (cement rendering to underside of tiles) and about 20 cracked pantiles.

Surface Water Drainage

Approximately one-half of the eaves guttering on the west side of building Nr 1 and one length on the south side of building Nr 3 are missing. All the gutters need cleaning out, especially the trough gutter on the east side of building Nr 1, which is causing dampness on the inner face of the east wall on both floors. The downpipes contain several cracked lengths and defective joints and are leaking. Gullies are badly silted and two gratings are broken.

Yard

Stone sett and concrete paving to yard is in bad condition; uneven and badly cracked.

Internal condition

Walls

The east, north and south walls of building Nr 1, the east wall of building Nr 2 and the west and north walls of building Nr 3 are damp. There is no evidence of any damp-proof courses. One large crack has occurred at the north end of the west wall to building Nr 1. All the walls need redecorating.

Concrete Floors

Rough surface with extensive cracks.

Boarded Floors

Sound condition, but rough finish and very unevenly laid with a number of small holes roughly patched. A space exists between the floor and the south wall of building Nr 3 for about half the length.

Roof Timbers

Fair condition. One strut in building Nr 2 is split.

Electrical Installation

Poor condition; the iron conduit is rusting and wiring to pendants is perished and in some cases missing.

Water supply

The one tap in the yard beside building Nr 2 is leaking and has no protection from frost.

The Surveyor as an Expert Witness

Function and Qualifications of Expert Witnesses

Mildred[3] has classified witnesses into two main categories: witnesses of fact (lay witnesses) and witnesses of opinion (expert witnesses). A witness of fact has a personal knowledge of events which happened in the past, and he is not normally permitted to express an opinion on the facts. A witness of opinion has special knowledge such as that acquired by a surveyor in the course of professional training and experience and he can assist a tribunal in coming to a decision by giving his opinion on the facts before the tribunal.

When, for instance, a building dispute arises, it may be settled by an action in the courts or by arbitration. The parties in dispute may employ legal advisers (counsel) who will prepare evidence and argue their case before the tribunal. On matters of a technical nature, the assistance of an expert witness may be required, and with building disputes this could be a surveyor. As indicated earlier, an expert witness is not restricted to giving statements of fact and can explain technical matters and also express an opinion, if requested, based on his special knowledge and experience.

Counsel will need to know the technical arguments and how they are to be presented and developed. Hence the expert witness prepares a document, termed a proof of evidence, for the benefit and use of counsel when arguing the case. The contents of the proof of evidence are copied into the counsel's brief and the expert witness will subsequently be questioned and cross-examined upon it.

The basic qualifications required of an expert witness are as follows:

(1) Relevant qualifications and experience in the area covered by the dispute.

(2) A general knowledge of the law of evidence, principles of damages, professional negligence, breach of contractual duty, and the duties of expert witnesses.
(3) The ability to communicate both orally and in writing in clear, simple language.
(4) The capacity to weigh facts and to draw logical conclusions from them.
(5) The ability to consider a problem impartially and with integrity.[3]

The expert must keep himself fully informed on the latest developments and practice in his field. He must, however, ensure that he does not hold himself out to possess special skills or knowledge beyond those which he does in fact possess, otherwise he may be liable in tort and in contract.

Fees of Expert Witnesses

Fees are best assessed on an appropriate time-rate charge agreed with the client in writing before any work is undertaken. Thereafter, all time spent on the assignment must be carefully recorded as it may subsequently be queried. Even where the arbitrator awards the client his claim, or confirms his defence against the claimant, and awards him costs, the client will not recover all the costs he has incurred. The expert witness's rate of payment and the time spent is normally taxed (reduced) on behalf of the losing party by the arbitrator or a taxing master of the High Court. In theory taxing should be no concern of the surveyor acting as an expert witness, whose financial terms of appointment were settled at the initial stage, but in practice problems can arise over payment and occasionally the surveyor may be obliged to sue for recovery of fees.[4]

Surveyor's Contribution as an Expert Witness

A surveyor may be called on to act as an expert witness in a variety of situations. The following examples serve to illustrate the wide range of disputes with which he may become involved.

(1) During the erection of a building which included excavation for a basement and underpinning work, slight settlement occurred which damaged the fabric of an adjoining shop owned by a third party. A dispute arose between the contractor and the shop owner regarding the liability for and the extent of the damage. A surveyor may be engaged to give evidence as to the extent of the damage and cost of the remedial work.
(2) Under a tenancy agreement the lease provided that 'the tenant shall keep and leave the premises in good tenantable repair. . . .repairs to the structure excepted'. A dispute has arisen concerning the extent of the repairs under the covenant at the termination of the lease and what constitutes 'the structure' and 'tenantable repair'. Surveyors might be engaged by both

parties to interpret these terms in the lease and to prepare schedules of dilapidations.

(3) A dispute has arisen between a landlord and tenant regarding the non-payment of rent and a counter-claim for loss of use of the premises. During a heavy storm, rainwater penetrated the roof and caused a large area of ceiling to collapse. Carpets and wall and ceiling decorations were badly damaged and the flat was stated to be uninhabitable. The tenant moved out and refused to pay the rent. The landlord is suing for the rent and the tenant is counter-claiming for loss of use of the premises and the cost of making good or replacing decorations, carpets and furniture. Surveyors could be engaged by both parties to give evidence on their behalf.

(4) An employer entered into a contract to pay £65 000 for a house to be erected on his land by a builder. The house was to be erected of good quality materials and in a thoroughly workmanlike manner in accordance with a plan and specification signed by the builder and the employer, and was to be complete in all respects and ready for occupation by a specified date. No architect was employed to superintend the work. The employer has paid £55 000 on account but refuses to pay more as he alleges that the work is incomplete. He bases his contention on a report by a surveyor who considers that the foundations are defective and that certain materials are of poor quality and do not comply with the specification. The builder contends that he has completed the house satisfactorily. A writ has been issued by the builder claiming the balance of the contract price. Surveyors are likely to be engaged to give evidence for both parties.

(5) A landowner submits a planning application to develop land adjoining a residential area with private houses. Planning consent is refused by the local authority. A surveyor may be engaged to prepare an appeal against the local authority's decision.

Evidence

Evidence has been defined as the testimony of witnesses and the production of documents and the like which may be used for the purpose of proof in legal proceedings. A fact is admissible if the law allows it to be proved by evidence. To be admissible a fact must be either in issue or have some degree of relevance to the facts in issue. Facts in issue are those facts which a plaintiff must prove in order to establish his claim and those facts which a defendant must prove in order to establish a defence, but only when the fact alleged by one party is not admitted by the other. Facts which are admitted, expressly or by implication, are not in issue.[3]

There are two kinds of evidence in civil hearings: evidence relating to facts and evidence relating to external opinions. Evidence of facts can be given by anyone, provided they are not mentally deficient, unwilling to testify on oath or affirmation, or too young to understand what they are doing. By comparison expert evidence should only be given by someone with the relevant technical knowledge and ample practical experience of the matters under investigation. In some cases the same person can act both as an ordinary and as an expert witness, such as where reference to the courts or arbitration arises from activities in which a professional was appointed to deal with technicalities before the dispute arose. Should the dispute not be concerned with his own expertise, he will merely give factual evidence, as an ordinary witness, proving the authenticity of documents he has handled and the like. Whether an ordinary or an expert witness, he should not express an opinion on another person's expertise.[4]

Instructions

A surveyor may receive instructions to act as an expert witness in a variety of different ways, including previous engagements, recommendation by a party to the dispute, reputation as an expert in the particular field or nomination by a professional society.

Mildred[3] has advised that the following matters should be considered by a potential witness before accepting instructions.

(1) He must be fully acquainted with all aspects of the dispute on which he will be required to act, and to be sure that they are within the scope of his personal knowledge, experience and competence.
(2) He must then decide whether or not the client will benefit from his opinion.
(3) Finally he must decide whether he can work within the prescribed timescale having regard to his other commitments.

If he then decides that he can act as an expert witness he should make a request to the client for instructions in writing. In his letter of acceptance, the surveyor, or other professional, will state his fees, normally expressed as an hourly rate. He will also need to cover such matters as travelling and hotel expenses, secretarial charges, any charges for laboratory or other tests, and whether or not the fee will be subject to Value Added Tax.[3]

Preparation of the Case

Initial procedure After acceptance of instructions, the expert witness should set about fully acquainting himself with all relevant information and documents. The documents can be quite extensive in a building dispute, encompassing such documents as contract conditions, specification, bill of quantities, contract drawings, supplementary drawings prepared during the course of the contract, architect's instructions, sub-contractors' and suppliers' quotations and accounts, daywork sheets, clerk of work's and foreman's diaries, records of hidden work, correspondence and the final account.

A meeting at the solicitor's office for a general discussion will help to clarify the matters in dispute and an exchange of views will help in determining the approach to be adopted. Much of the expert's preliminary opinion may be based on assumption and/or deduction, and requires clarification by reference to documents known to be in the possession of the opposing party or of third parties. The procedure known as 'discovery of documents' needs to be implemented, whereby a judge or arbitrator can order the production of information by one party to the other if he is satisfied that it is necessary for full consideration of the matters in dispute.

The expert may consider that his instructions are too tightly drawn or need amending and, if this is the case, he should endeavour to get them changed in agreement with the instructing solicitors. The expert should also contact any other experts acting for the client to establish their approach and to avoid any possibility of overlap. A site inspection is often advisable at which the expert will make full notes of relevant matters and take photographs as necessary. There is nothing more damaging to an expert's evidence than to have to admit, in answer to a question in cross-examination, that he has not seen the site. It is preferably seen at least twice, before and after the preparation of the draft proof of evidence.

Subsequent procedure The next step is often the preparation of a report, encompassing such matters as the terms of reference, list of documents examined, inspections and facts noted from them and the expert's conclusions.

It is also advisable to consider the dispute from the viewpoint of the other party, and to try to anticipate the main points on which he is likely to rely and to devise convincing refutations to them. Where there are issues which go to the root of the dispute and cannot be answered satisfactorily, the expert would be wise to advise the client, through the instructing solicitor and counsel, to settle before the hearing.[4]

All weaknesses in the client's case must be identified. For example, an expert may be engaged to support a contractor's claim for direct loss and expense arising from delays in completion. The main heads of claim normally include the effect of variations ordered by the architect and the failure of the architect to provide information at the appropriate time. The expert must, however, also consider possible counter-arguments, such as changes in supervisory staff causing disruption of the work, shortages of labour and material, adverse effects on the programme of remedying defective work and disruption caused by sub-contractors leaving the site because of delays in payment by the contractor.

After the expert has submitted his report, it is customary to arrange a conference with the instructing solicitor and counsel to clarify any points that are obscure or are dealt with inadequately and to determine strategy.

The expert may be instructed to reach agreement with the opposing party on various matters in order to narrow the area under investigation. In a construction dispute ascertainable facts could include the form of construction and leading dimensions, material prices and wage rates. Questions involving professional judgement could include the cost of making good defective work (disregarding the reason why it was defective) and the extent of incomplete work (regardless of why it was incomplete). Mildred[3] recommends that the agreed documents should be endorsed with suitable wording, such as 'These facts and figures are agreed entirely without prejudice in regard to liability on the matters in dispute between the parties', and the endorsement signed by both parties.

Preparation of documents   The investigation may involve the consideration of numerous documents and these are best categorised in bundles and listed in an index for ease of reference. Each bundle may be allocated a reference letter and each separate item in the bundle numbered sequentially. Four sets of documents are normally required – one for the tribunal, one for witnesses and one for each party.

   Maps and plans should be neatly folded and referenced, both inside and out. Photographs should preferably have a label showing on the front face the details of the subject and the date when it was taken. Where a number of photographs are to be presented, they are best bound in book form with a fold-out index and a site plan showing where the photographs were taken.

Scott schedules   A Scott schedule has proved to be a logical and useful method of detailing the numerous items in issue in construction disputes. The schedule often reduces significantly the scope of the dispute as once they appear in schedule form the parties frequently agree quite quickly a large number of the smaller items. Typical examples of entries in Scott schedules are given in tables 6.2 and 6.3. In cases heard before an Official Referee, the headings claimant, respondent and arbitrator would be replaced by plaintiff, defendant and Official Referee. A further example of a Scott schedule appears at the end of chapter 8.

Planning Appeals

The client working closely with his consultant will probably decide to appeal against the planning authority's refusal to grant planning permission, but at the same time attempt to negotiate a solution with the planning authority. The consultant thus has two assignments.

(1) He should draft the grounds of appeal for discussion and agreement with the solicitor or counsel and then lodge the appeal and its related documents.
(2) The consultant will open up discussions with the planning authority's officers in an attempt to negotiate an acceptable compromise. He will be prepared to approach these discussions with a fresh mind and possibly new ideas, and hopefully he will be able to convince the planning authority's officers that the revised scheme is based on sounder planning principles and produces a greater measure of planning gain than the earlier rejected scheme.[5]

Table 6.2 Claim by contractor for extras

Item Nr	Reference	Description	Claimant's comments	Claimant's price	Respondent's comments	Respondent's price	Arbitrator's comments	Arbitrator's sum awarded
23	Architect's instruction 34(b)	Use of 66 m^3 of ready mixed concrete instead of site mixed concrete	Needed for construction of extension authorised after removal of batching plant	£198	Ready mixed concrete should not normally exceed the cost of site mixed concrete by more than £1 per m^3	£66		

Table 6.3 Items in dispute on a remeasurement contract

Item Nr	Reference in Schedule of Rates	Description	Claimant's				Respondent's				Arbitrator's	
			Comments	Quantity	Unit rate	Sum claimed	Comments	Quantity	Unit rate	Sum offered	Comments	Sum awarded
					£							
15	28/D	Substitution of 100 diameter clay flexible jointed pipes laid and jointed in trench, for clay pipes with cement mortar (1:3) joints	More expensive pipes than original	215 m	4.90 3.80 ——— £1.10 ———	£236.50	Pipes are more expensive but laying and jointing is much cheaper, so the net difference should not exceed 15p per m	215 m	15p	£32.25		

The consultant will advise on whether any additional expert evidence is needed and will co-ordinate and reconcile all the evidence in the form of draft proofs of evidence. The consultant as principal witness, working closely with the solicitor, counsel and other expert witnesses, will aim to produce a sound and well-considered case.

The Hearing

General procedure Procedure at a hearing varies according to the type of tribunal hearing the case. Court cases follow a prescribed procedure and arbitrations are similar but can be more flexible. The hearing commences by the counsel for the plaintiff (claimant in arbitration proceedings) reading the pleadings which constitute largely the points of claim against the defendant (respondent in arbitration proceedings). He then calls his first witness and takes him though his proof of evidence in question and answer form; this is termed examination-in-chief. With few exceptions, leading questions which anticipate the witness's reply are not permitted in examination-in-chief. For example, a question phrased 'When you inspected the property, dampness was penetrating the walls, wasn't it?' is not permissible. The question could however be put as 'What was the state of the walls when you inspected the property?' The exceptions to the rule about leading questions relate to the identification, qualifications and experience of a witness, and the identification of documents.

The advantages of using a proof of evidence in examination-in-chief is that the witness does not have to rely on his memory, resulting in greater reliability and reducing the time taken in examination. An expert witness may refer to standard works and recognised authorities, in addition to reference to notes made by him at the time when the events occurred.[3]

On conclusion of examination-in-chief, counsel for the defendant cross-examines the witness. The purpose of cross-examination is to test the truth of a witness and the accuracy and completeness of his evidence, with the prime objective of weakening or even destroying the opposing case. Counsel is not obliged to cross-examine, but evidence given by an expert witness is likely to be controversial and hence he can anticipate being cross-examined. Leading questions are permitted in cross-examination, but counsel must take care not to impugn the character of an expert witness.

When the counsel for the defendant has completed the cross-examination, the counsel for the plaintiff may re-examine the witness, by asking further questions to elucidate points raised in cross-examination. No new matters may be introduced at this stage.

The remaining witnesses for the plaintiff are all dealt with in the same manner, but their sequence is a matter for counsel.

Counsel for the defendant may make an opening address or proceed forthwith to the examination of witnesses, working through the same processes of examination-in-chief, cross-examination and re-examination. Finally, counsel for the defendant makes his closing address followed by the closing address for the plaintiff. The judge in a case not heard by a jury will either give his judgement immediately or reserve it and deliver it subsequently on a named day. In the case of an arbitration, the arbitrator normally reserves his judgement which is contained in a written document known as the award.[3]

An expert witness will probably be present only for part of the hearing, to give his own evidence and to hear that of the expert(s) for the opposing party. In this way he can advise counsel through the instructing solicitor on points raised during the examination of the opposing witnesses. Attendance throughout the whole of the hearing would be costly and serve little purpose.

Procedure at public inquiries Public inquiries are held before inspectors employed by the Department of the Environment in connection with appeals against decisions on planning applications, building regulation applications, highway and sewerage proposals and related matters. The procedure at public inquiries is similar to that adopted in a Court of Law, but a greater measure of flexibility and informality is permitted. There is, for instance, some relaxation of the rules of evidence, and it is customary for witnesses to read their evidence after circulating copies to the inspector and the opposing party. The procedure at a public inquiry dealing with an appeal on a planning or building application is as follows:

(1) The appellant or his representative makes an opening statement outlining his case.
(2) The appellant calls witnesses to give evidence and these are cross-examined by the local authority's representative and re-examined, if necessary, by the appellant. The Ministry inspector may also ask questions if he wishes to do so.
(3) The local authority's representative makes an opening statement and calls his witnesses in the same manner as the appellant.
(4) The local authority's representative can make a closing statement if he so wishes.
(5) The appellant or his representative makes a closing statement.
(6) Any other interested parties are invited to make statements and may be cross-examined.
(7) The Ministry inspector visits the site accompanied by a representative of each party, who can identify objects mentioned at the inquiry.

The Ministry inspector prepares a report for submission to the Minister, who subsequently communicates his decision in writing to the parties, with reasons for his decision.

Expert witness's approach   The expert witness should approach the witness table in a relaxed manner but give evidence with decisiveness and confidence. He should speak clearly with even presentation and at a reasonable speed. The person conducting the proceedings will take notes in manuscript and the witness should adjust the pace of his presentation to suit the time required for note taking. The witness should address both the tribunal and the questioning counsel.

Questions should be answered clearly and concisely, avoiding verbosity and repetition. Technical jargon should be avoided wherever possible, keeping statements as simple as possible. The witness will need to use his powers of concentration to the utmost, and to be prepared to give reasoned answers to the most provocative questions put by opposing counsel.

Proofs of Evidence

Contents of Proof of Evidence

A proof of evidence in some ways resembles a report and must convey the facts and opinions of the expert witness clearly, simply and succinctly. Whereas a report is prepared for a lay client, a proof is written for counsel and used by him when leading his expert witness. Whether the proof of evidence covers a rating appeal, a planning inquiry or a building dispute, the general approach will not vary a great deal. The contents of the proof generally contain the following components.

(1) *Title*: this can be given on a separate title page or clearly stated as a main heading on the first page.
(2) *Name and qualifications*: the expert witness's personal particulars, incorporating his qualifications, period in practice with positions held during this period and details of experience in the particular aspects of practice with which the reference is concerned.
(3) *Subject of reference*: a brief recital of the subject matter and any relevant facts such as dates of specific events, notices and the like.
(4) *Site inspection*: dates and details of site inspections.
(5) *Evidence*: this will form the main part of the proof and will give details of the expert's findings and present and develop arguments. Exhibits such as detailed estimates of cost may be included in the body of the proof or be attached to it. Each statement must be capable of verification and all calculations must be carefully checked. Hearsay evidence supplied by third parties which is not capable of verification by the expert must be excluded.
(6) *Conclusions*: opinions based on the expert's investigation of the facts and particulars of other evidence, such as reference books, professional and technical papers and the like, which support his conclusions.

Format of Proof of Evidence

The normal format of a proof of evidence is now described.

(1) The document should be typed in double spacing on brief size paper which is equivalent to A3 international size (410 mm x 295 mm).

(2) Wide margins should be left for notes, additions or corrections inserted by counsel.

(3) Paragraphs should be referenced by numbering. Pages should also be numbered and adequate headings and sub-headings inserted.

(4) Valuations, schedules and other exhibits may be dealt with either by reproducing them on A3 paper and securing them at the end of the proof or, if they consist of only a few lines, including them in the body of the proof at an appropriate place in the text. In the latter case the exhibit is indicated by square brackets, a marginal line or a reference number in the margin.

(5) Words of special importance may be underlined to catch the eye of counsel.

Examples of Proofs of Evidence

Two examples of proofs of evidence are now given, one relating to building defects in a recently erected dwelling and the other to evidence to be given at a public inquiry concerning a compulsory purchase order and clearance of a residential area. A further example, covering building repairs at the termination of a lease, is included in chapter 8.

Building Defects to 38 Woodrush Close, High Norton
PROOF OF EVIDENCE
OF
JOHN WILSON PETERS

My name is John Wilson Peters.

I am a Fellow of the Royal Institution of Chartered Surveyors and a Senior Partner in the practice of Building Surveying Partnership of 96 High Street, Blandworth. Building Surveying Partnership has a wide experience of the design. conversion, refurbishment and maintenance of all types of buildings, and I have been involved with all of these activities during my 24 years with the practice.

I was instructed by Wilkins and Jones, Solicitors, of 18 High Street, Blandworth, to investigate alleged building defects at 38 Woodrush Close, High Norton, and to prepare a report on the defects and a claim for their rectification.

Background to claim

1.1 The property is a two-storey dwelling house erected by Homebuilders Ltd for Joseph Smithson and completed on 9 July 1983. Joseph Smithson engaged an architect, Peter Ramsay, to design the house, and then contracted with Best Homebuilders Ltd to build the house in accordance with the plan and specification prepared by the architect at a contract price of £48 200. The architect was not engaged to supervise the construction.

1.2 In March 1985, the owner observed dampness on the internal face of the south-west wall in the lounge and also dampness and some timber decay in the north-east corner of the roof space. He contacted the builder who refuted any liability. Best Homebuilders Ltd are not a member of the National House-Building Council.

Inspection

2.1 I was supplied with a copy of the plan and specification and I inspected the property on 15 May 1985.

2.2 I found six damp patches on the south-west wall of the lounge. There had been a period of fairly steady rain during the two days preceding my visit. Using a metal detector, I was able to establish that the damp patches were in every case close to a wall tie in the cavity of the external hollow wall. One facing brick was cut out adjacent to a damp patch and showed substantial mortar droppings on the exposed wall tie.

2.3 An examination of the roof space in the low pitched roof revealed a high level of condensation, particularly in the north-east corner. Impermeable sarking felt has been provided under the roofing tiles and a 4" (100) layer of glass fibre insulation has been laid between the ceiling joists and finishes tight against the eaves, thus effectively sealing the roof space. Moist air is gaining access to the roof space through the access hatch which is loose fitting and through holes in the ceiling around service pipes. A badly fitting cover to the cold water storage tank is permitting water to evaporate into the roof space. There is no chimney stack passing through the roof space. Six rafters and ceiling joists are decaying as a result of an attack of wet rot.[6]

Causes of Defects

3.1 The dampness on the south-west wall of the lounge is due to mortar droppings on the wall ties, which permit rainwater penetrating the outer brick skin to bridge the cavity and enter the inner block skin. Good building practice is to remove daily any mortar which inadvertently falls on the wall ties, while bricks and blocks are being laid. The concentration of dampness on the south-west wall results from the prevailing moist wind being from the south-west, so that this wall is the most vulnerable from the damp penetration aspect.

3.2 In my opinion the outbreak of decay in the roof timbers results mainly from the excessive level of condensation in the roof space and the absence of any form of ventilation. The additional precaution of treating the timbers with suitable preservative has not been carried out.

Conclusions

4.1 The architect's specification does not specifically require the builder to keep all wall ties clear of mortar droppings. Nevertheless this is generally accepted in the building industry as good practice. It is highlighted in British Standard Code of Practice *CP 121, Part 1: 1973, Walling: Brick and Block Masonry*, which constitutes an authoritative document detailing good practice.

4.2 Neither the architect's specification nor the plan make any reference to ventilation of the roof space. Here again, good building practice entails the provision of positive ventilation. A minimum of a $\frac{3}{8}''$ (10) gap should be provided between the edge of the insulation and the eaves. In addition adequate ventilation holes should be formed in the fascia and/or soffit board at the eaves. A Building Research Establishment recommendation is to provide at least ¾ in^2 (300 mm^2) free opening for every 1 ft (300 mm) run of eaves. These recommendations are detailed in Building Research Establishment *Digest 180: Condensation in Roofs* (HMSO, 1975) and *Digest 218: Cavity Barriers and Ventilation in Flat and Low Pitched Roofs* (HMSO, 1978).

4.3 The remedial work involves cutting out facing bricks from the outer skin from points just above the areas of damp penetration, removing mortar and any other obstructions in the cavity and then making good the wall with new matching bricks. After the damp wall has dried out it will be necessary to redecorate the walls to the lounge with two coats of emulsion paint.

The decaying roof timbers must be cut out and replaced with sound, well-seasoned and treated timbers. A gap $\frac{3}{8}''$ (10) wide must be formed between the ceiling insulation and the eaves and holes formed in the eaves soffit boarding to provide ¾ in^2 (300 mm^2) of free opening every 1 ft (300 mm) run of eaves.

4.4 It is my considered opinion that the builder is liable for the defects that have occurred in the walling and roof to the house, as he failed to carry out established building practice which the building owner was entitled to expect. The failure of the architect to specifically detail these matters is not relevant to the issue and this view is supported in the decision in *Brunswick Construction Ltee v. Nowlan*, Supreme Court of Canada (1974), 21 BLR 27. In any building contract, unless the parties have agreed to the contrary, the builder impliedly agrees:

(1) that he will do his work in a good and workmanlike manner;
(2) that he will supply good and proper materials; and
(3) (where the contract is to build a house) that the house will be reasonably fit for human habitation.[7]

There are two options available for the rectification of the defects:

(1) for Best Homebuilders Ltd to carry out the work that is detailed in this proof; or
(2) to authorise another builder to undertake the work and for Best Homebuilders Ltd to pay his costs.

The estimated cost of the remedial work is £650.

J. W. Peters
22 May 1985

Representations to be made at a Public Inquiry
Mottlesham – Stansted Clearance Compulsory Purchase Order
Public Inquiry, 6 June 1985
PROOF OF EVIDENCE
OF
CHARLES BECKETT

My name is Charles Beckett

I am a Fellow of the Royal Institution of Chartered Surveyors and a Fellow of the Royal Town Planning Institute. I am a partner in Johnson and Smith, Surveyors and Planning Consultants of 21 Queen Street, Mottlesham. The practice has gained a wide experience over the last 30 years in planning, development and housing work and for the past 12 years I have specialised in the redevelopment and rehabilitation of housing areas, and prior to that I was employed for 8 years as a planning assistant in a city planning office.

My practice was instructed by the Stansted Area Residents' Association to represent them at the public inquiry and to put forward their objections to the clearance proposals.

Clearance Proposals

1.1 The clearance proposals prepared by Mottlesham City Council provide for the large scale clearance of 1254 houses in the Stansted Area of the City and their replacement by approximately 230 dwellings, together with the provision of some industrial units, a school, some open space and various community facilities. The proposals if implemented must therefore result in the movement of a very large number of residents from the clearance area.

Need for Greater Selectivity

2.1 At the outset I wish to make it clear that a considerable number of the members of the Residents' Association welcome the clearance proposals *in principle*. It is, however, debatable whether individual houses are fit or unfit, although it must be conceded that a significant number of properties are in appalling condition. These houses are a threat to the health and well-being of their occupants. For example, the dwellings in Blenheim Terrace probably number among the worst in the entire city. For many residents the opportunity of being rehoused in modern local authority accommodation cannot come too soon. They are keen to move out and should ideally be moved as soon as possible.

2.2 However, it is important to bear in mind that the people living in the area are no more identical than the types and condition of the houses that they occupy. Some are young families with children, living in cramped and unhealthy conditions. Others are elderly persons, who have lived in the area for 30 years or more and who have spent a lot of time, money and care in making their homes pleasant places in which to live.

2.3 The purpose of the inquiry, as I understand it, is firstly to determine whether individual houses are unfit, and secondly to decide whether the most satisfactory way of dealing with the conditions in the area is to demolish all the properties covered by the compulsory purchase order as it now stands.

2.4 Thus there is a *diversity* of housing conditions and housing needs in the area, and no *uniform* treatment, such as large scale demolition and redevelopment as proposed, can adequately and sensitively relieve these conditions or meet these needs.

2.5 The compulsory purchase order plan shows a scatter of grey and pink, fit and unfit dwellings. In some parts, such as around Stansted Street itself, there are apparently blocks of unfit houses and, in these cases, demolition may be the only feasible solution. There are, however 265 properties shown as fit; some of these are located individually, or in groups of two or three, but others form complete terraces. Numbers 51 to 77 and 93 to 109 Haslemere Road, 125 to 139 Tennyson Avenue, 12 to 32 Walton Street, and 12 to 28 and 60 to 82 Cowbridge Road are, among others, shown as fit. These comprise six terraces containing a total of nearly 60 fit houses, providing homes for possibly 140 to 180 people, and the City Council proposes to demolish them all.

2.6 I understand from the City Planning Office that one area of fit houses around St John's Street is likely to be removed from the clearance compulsory purchase order and the Association is delighted to hear this. The City Council has in fact admitted that these houses, which it originally considered had to be demolished to provide a satisfactory site for redevelopment, can now be retained. It came to this conclusion after repeated requests at a public meeting, the presentation of petitions to the Chairman of the Housing Committee, and the voicing of opinions by those residents who were able to visit the Information Bureau in Radstock Road, which unfortunately was closed in the evenings.

2.7 I would like to pose the question – If the City Council's officers, with all their technical expertise, have been mistaken once, might they not be mistaken in other instances? Might it be practicable to retain other fit properties in the area, and improve them to an equally high standard as that which the City Council hopes to achieve in St John's Street.

2.8 I submit that it is unreasonable to state that certain properties must be cleared in order to provide a satisfactory site for redevelopment, unless attempts at designing a redevelopment scheme retaining such properties have been made and have failed. It is only by experimenting with alternative layouts, some of which retain fit houses, that it is possible to state categorically that it is impracticable to redevelop without demolishing them.

2.9 There are, of course, different degrees of fitness and unfitness. I consider that it is particularly important to determine whether houses which are marginally below the threshold of fitness are capable of being made fit at reasonable expense. Opinions will vary as to what is 'reasonable' in this context as, for instance, some local housing associations are carrying out work to render unfit houses fit at an expense which the local authority considers prohibitive.

2.10 It may be argued that some of the fit terraces that the Association wish to see retained are too cramped by modern standards and that, in consequence, they should be cleared. I would like to respond by saying that the fact that a dwelling is small does not necessarily make it incapable of improvement. I reject the criterion of lack of space as an adequate measure of housing unsuitability, particularly since there are many terraces in existing and proposed General Improvement Areas which are, strictly speaking, sub-standard in terms of internal and external space and yet provide adequate homes for hundreds of families.

Social Effects

3.1 The population of the Stansted Area is relatively stable. Many people have lived there for more than 20 years, and a large proportion of the elderly have been residents for 40 years or more. The 1981 census showed that more than a third of the households in the area contained at least one pensioner. Elderly people are very dependent on nearby friends and relatives for help with shopping, cleaning, cooking and psychological support, and are very disturbed at the thought of having to leave the area.

3.2 For those people who have managed to buy a house, the clearance plan means that they will probably never have the chance to be owner occupiers again. They can no longer look forward, as many did previously, to selling their present home in a few years' time and buying a better one with the proceeds. No amount of compensation can make up for the loss of a home of one's own.

3.3 Many middle-aged and younger families living in the area work in the city centre, which is readily accessible by public transport or they can walk or cycle to work.

3.4 Partly as a result of the clearance proposals, people have stopped caring and houses have become more dilapidated over the past 18 months. In this way a proposal for large scale demolition becomes a self-fulfilling prophecy.

3.5 Had a more constructive and imaginative approach been adopted, embracing the retaining of reasonable houses, repairing marginal ones and selectively replacing the worst, wholesale blight could have been avoided and, along with it, a good deal of human misery.

3.6 I have detected an undercurrent of dissatisfaction with the way in which the area has been allowed, even encouraged, to run down. Many people seem unable to express their feelings publicly; they do not understand the legal procedures and technical jargon surrounding the serving of the compulsory purchase order and the making of objections to it.

3.7 Many working people cannot afford the cost of presenting a case at a public inquiry, an activity which the local authority undertakes at the ratepayers' expense. Neither can all those residents who wish to attend find the time to do so.

3.8 In this context, no one can claim that the people have participated in the preparation of the plan for the Stansted Area. Other than holding two public meetings, the City Council has made little attempt to encourage residents to feel that they have any share in deciding the fate of their homes and the future of their area. Despite public undertakings to arrange for full scale participation, the City Council has not even carried out a social survey. Hence many residents have adopted a fatalistic attitude towards the whole scheme.

Conclusions

4.1 I share deeply the concern of the Residents' Association at the way in which the

City Council has approached the redevelopment of the Stansted Area. There has been only minimal consultation with the residents who are so deeply affected by the proposals. Greater involvement of the residents in the early stages of formulation of the proposals would have reduced substantially the present dissatisfaction and could have resulted in more balanced and acceptable proposals.

4.2 I am disturbed at the unimaginative one-sided proposals which entail such a large scale destruction of property, regardless of its condition and the needs and wishes of residents. The scheme appears to have been prepared with ease and convenience of implementation as the first priority, and shows a serious lack of sensitivity, selectivity and understanding.

4.3 Apart from the failure to make maximum use of existing housing stock, the proposals disregard the advice in the very relevant Government White Paper – *Better Homes: the Next Priorities* (1973)[8] which placed the emphasis on improvement of houses and the operation of Housing Action Areas as opposed to large scale clearance and redevelopment. Indeed the present proposals appear extremely outdated as government and local authority policy generally has, since the late nineteen sixties, been concentrating on rehabilitation.

The White Paper expresses the residents' sentiments and aspirations admirably and I believe that it would be helpful to quote some relevant extracts from the White Paper – "... in the majority of cases it is no longer preferable to attempt to solve the problems arising from bad housing by schemes of widespread, comprehensive redevelopment. Such an approach often involves massive and unacceptable disruption of communities and leaves vast areas of our cities standing derelict. ... Increasing local opposition to redevelopment proposals is largely attributable to people's understandable preference for the familiar and, in many ways, more convenient environment in which they have lived for years. ... The Government wishes to encourage the concept of gradual renewal which allows groups of the worst houses to be cleared and redeveloped quickly; some to be given minor improvement and repair pending clearance in the medium term; others comprising predominantly sounder houses to be substantially rehabilitated and possibly be included in General Improvement Areas. This approach also means that fewer houses, at any point in time during the process of renewal, are lost from the stock of available dwellings – a very important consideration in areas of high scarcity."

4.4 I strongly urge that the approach recommended in the White Paper be implemented for the Stansted area in full consultation with the Residents' Association. A primary aim should be to retain and modernise as many fit and marginally fit houses as possible, compatible with producing a convenient and attractive scheme, as a more economic and practical solution and one causing less social upheaval than the City Council's proposals. The inclusion of some additional amenities and social facilities is to be welcomed, but the scale and scope of this provision should be kept within reasonable limits, having regard to the restricted physical area available and consequent loss of housing, and the possibility of providing some of the facilities on the periphery in neighbouring areas. The intention to clear the area in a single operation should, I believe, be abandoned in favour of a more gradual programme of rehabilitation and redevelopment.

Charles Beckett
11 May 1985

Arbitration

General Background

Arbitration is a formal process for the settlement of disputes, provided for in all the standard forms of construction contract and sub-contract, as an alternative to court proceedings. The basis of arbitration is that the parties to a dispute select a person on whose judgement they are prepared to rely, and agree to abide by the decision that is reached. A number of professional bodies, including the RIBA and RICS, maintain panels of arbitrators, and will at the request of the parties appoint a suitable person to act as arbitrator. Every arbitration, unless the parties express a contrary intention, will be controlled by the provisions of the *Arbitration Acts of 1950 and 1979.*
    The advantages claimed for reference to arbitration are:

(1) the process is generally a voluntary one;
(2) the proceedings are conducted in private;
(3) the decision is final;
(4) it is generally believed that the process is quicker and less costly than a court hearing, although some arbitrations have lasted for years with mounting costs (furthermore, the arbitrator has to be paid and a room hired for the hearing, whereas in the High Court, the judge and court are free); and
(5) the person selected as arbitrator is often an expert in the matter under dispute.[9]

Arbitration Procedure

Procedure in arbitration follows very closely the procedure in the courts, for the same rules apply. After the arbitrator is appointed he will first hold a preliminary meeting, attended either by the parties (claimant and respondent) or by their solicitors, at which he will become familiar with the nature of the dispute. At this meeting he will give directions as to the conduct of the case and its timing, including:

(i)   preparation of the claimant's case (points of claim);
(ii) respondent's answer to the claimant (points of defence) and the counterclaim if the respondent feels that he has a claim against the claimant; and

(iii) claimant's reply to the defence, and a defence against the counter-claim where there is one.[10]

These three sets of documents are referred to collectively as the pleadings. When they are complete a procedure termed 'discovery' is initiated. It is a requirement of English law that any documents, such as correspondence and drawings, held by either party, which touch on the matters in dispute and which may be used in evidence shall be disclosed to the other party. The hearing is conducted in a similar manner to that in the courts as described earlier in the chapter. Subsequently the arbitrator makes his award and it is delivered to the parties. There is no appeal against the award of an arbitrator and the courts will enforce a valid award.

References

1. I. H. Seeley. *Building Technology*. Macmillan (1980)
2. H. S. Staveley and P. V. Glover. *Surveying Buildings*. Butterworths (1983)
3. R. H. Mildred. *The Expert Witness*. Godwin (1982)
4. W. James. The expert witness. *Quantity Surveyor* (November 1974)
5. R. Mercer. The role of the expert witness. *Chartered Surveyor* (January 1981)
6. W. H. Ransom. *Building Failures*. Spon (1981)
7. V. Powell-Smith and M. Furmiston. *A Building Contract Casebook*. Granada (1984)
8. White Paper. *Better Homes: the Next Priorities*. HMSO (1973)
9. J. Parris. *Arbitration: Principles and Practice*. Granada (1983)
10. A. B. Waters. Arbitration in building disputes. *The Practice of Site Management*. Chartered Institute of Building (1980)

7 Dilapidations

This chapter is concerned with the nature, significance and application of dilapidations. A general understanding of the implications of repair clauses in leases, relevant statutory provisions and the liabilities of the owners and occupiers of property is required. The situation with regard to such matters as fences, fixtures and party walls is also considered.

Nature of Dilapidations

The term 'dilapidations' refers to the disrepair or dilapidated condition of land or buildings, in situations where a legal liability is imposed upon the person(s) responsible. The person whose acts of omission or commission have caused the dilapidations is normally one with a limited interest in the property, such as a tenant for life or a lessee under a lease, whose neglect to keep the property in a good state of repair will have detrimental consequences for those who take possession of the property when his interest terminates.

Most dilapidations with which the surveyor is concerned form a part of civil law. Another category of dilapidations is ecclesiastical, where the incumbent of a benefice allows the parsonage house or outhouses to fall into disrepair and the relevant law is found mainly in the *Repair of Benefice Buildings Measure 1972.*

Waste

Civil dilapidations extend beyond the law of landlord and tenant to embrace the law of waste. Waste has been defined as spoil or destruction to houses, gardens, trees or other corporal hereditaments, to the injury or detriment of the reversion of the property, and the two types of waste of greatest practical importance are voluntary waste and permissive waste.

Waste has also been described as unauthorised acts by a tenant which alter the character of a property. The law of waste aims at preventing a tenant from treating the property in such a way as to reduce or restrict the rights of those who are to succeed him. Waste is a tort and, as such, requires proof of actual damage. It cannot be assigned and is essentially a legal wrong exclusive of breach of contract.[1]

Four types of waste are identifiable and these are now described.

(1) *Voluntary waste* consists of any unauthorised act which alters the nature of the property and injures the inheritance. Typical examples are pulling down buildings, destroying or removing fixtures, felling timber trees or changing the form of husbandry, such as converting arable land into pasture or vice versa.

(2) *Ameliorating waste* is a form of voluntary waste which improves the land or enhances its value, resulting in betterment of the reversionary interest in the land. There can be no remedy at common law because no damages can be assessed and equity would refuse to grant an injunction to restrain the doing of an act of waste which would cause no injury to the reversion. Typical examples are converting unused or derelict buildings into operational buildings or waste land into building plots.

(3) *Permissive waste* is the failure to repair the property and to maintain it to an acceptable standard, possibly even allowing it to fall into a state of ruin or neglect. West[1] describes how the liability for permissive waste is imposed mainly on tenants for years and is concerned primarily with damage to buildings resulting from unauthorised and unreasonable neglect to carry out such repairs as are necessary to preserve the structure from premature decay. Such neglect normally constitutes a breach of a lessee's express or implied contract to repair, and the remedy is accordingly an action for damages for breach of contract.

(4) *Equitable waste* arises where an occupier of land or buildings, such as a tenant for life or years, is exempt from his normal liability for waste in the instrument creating his estate. He is thus declared to be 'unimpeachable of waste' or 'without impeachment of waste'. Thus if he abused his powers and committed acts of wanton destruction such as stripping the house of lead, windows and other fixtures, pulling down the principal buildings or cutting down ornamental trees, common law provided no remedy.

However, courts of equity could have intervened to protect the interests of successors, by granting an injunction to restrain this abuse of power and hence the use of the term 'equitable waste'. The matter has also been the subject of legislation – the *Law of Property Act 1925, section 135* states: 'An equitable interest for life without impeachment of waste does not confer upon the tenant for life any right to commit waste of the description known as equitable waste, unless an intention to confer such right expressly appears by the instrument creating such equitable interest.'

Irrespective of the doctrine of waste, the destruction of buildings of special architectural or historic interest is forbidden by the *Town and Country Planning Act 1971.*

Leases

The relationship between landlord and tenant is governed by a contract of tenancy, frequently termed a lease, whereby the landlord allows the tenant to have exclusive occupation and quiet enjoyment of the landlord's house or land for a specific period, and in return the landlord receives a periodic payment of money called rent. Leases vary in their form and implications according to the length of term which is granted.

The longest term leases are usually *building leases*, by means of which a landlord usually lets to a tenant a piece of vacant land for a term of say 99 or even 999 years at a ground rent, being a rent for the land alone. The tenant agrees to erect a house or other property on the land of a certain value. The landlord takes the ground rent and ensures that the tenant performs his repairing obligations. When the lease ends, the land and building revert to the landlord.

Occupation leases are leases of fully improved property at rack rents, these being full market rents subject to statutory controls, approximating to the full improved annual value of the property. In the case of specific lettings for periods of 7, 14 or 21 years or longer, the tenant usually undertakes responsibility for all repairs and pays all other outgoings.

Lettings for terms of three years or less and tenancies from year to year may be evidenced merely by the exchange of letters between the parties. The tenant seldom agrees to accept responsibility for repairs beyond making good damage caused by him and the payment of tenant's rates and taxes.

Other variations in the form of leases include the granting of a lease for a *fixed period* of specific and certain duration. Another type is known as a *yearly tenancy* which continues from year to year indefinitely, until it is terminated by proper notice. Yearly tenancies may be implied in certain cases as, for example, where a person with the landlord's consent occupies property and rent is assessed, paid and accepted on a yearly basis. *Periodic tenancies* are created in a similar manner to yearly tenancies but operate for shorter periods, such as week to week, month to month or quarter to quarter. *Tenancies at will* arise when either the landlord or the tenant may end the tenancy at any time. A *tenancy at sufferance* operates when a tenant who was in lawful possession of property continues in possession after his right to occupy has expired and without the consent of the landlord.

Covenants

A lease usually contains a number of terms and conditions agreed by the parties. Before the lease is drawn up, the parties normally enter into a contract whereby the landlord undertakes to grant a lease of the specified premises to the tenant. The terms and conditions to be observed may be stated in the contract, but this is rarely so in practice. In these circumstances certain covenants will be implied and are often referred to as 'the usual covenants'. When the lease is finally drawn up it will contain these covenants in full, and they normally include undertakings by the tenant to keep and deliver up the premises in repair and to allow the landlord to enter and view the state of repair.

A covenant is in essence a promise contained in a deed. In general, covenants are governed by the law of contract whereby only parties to the deed can enforce the covenants contained within it. However, this is not the case with transactions relating to land. For example, A may be bound by covenants made by B in an agreement with C, resulting in A being bound by covenants arising out of a transaction to which he was not a party.

When the landlord and tenant enter into a contractual relationship they are bound by privity of contract and each of them can compel the observance of the covenants in the lease. If the landlord assigns the reversion or the tenant assigns the lease, each is still liable under the covenants, and in the event of the death of either party their liabilities will devolve upon their personal representatives.

The covenants which can be enforced on the assignment of a lease must 'touch and concern' the leased property. Thus covenants by a lessee to pay the rent, carry out repairs, insure against fire, or use the property only for a dwelling house, are directly concerned with or touch and concern the property. In like manner covenants by the lessor to renew the lease, carry out external repairs, or not to build on adjoining land can be similarly classified.

Liability for Repairs

Condition of Premises at Commencement of Tenancy

Unless the landlord has expressly agreed to do so, or statutes such as the Housing Acts have imposed the liability on him, he is under no obligation to put the premises into repair at the commencement of the tenancy or to carry out repairs during the tenancy. As expounded by West,[1] at common law, in the case of a contract for letting land or buildings or an unfurnished house, there is no implied condition or stipulation that the premises are fit for habitation or occupation, or cultivation in the case of land. It is for the tenant to satisfy himself that the premises are suitable for his purpose as he takes them as they stand with all their benefits and deficiencies.

It is accordingly important, in the absence of an express warranty by the landlord concerning the condition of an unfurnished house, for the intending tenant to have the property carefully surveyed by a competent professional, even where the tenant does not undertake to keep the premises in repair. If the tenant fails to take this precaution he could incur serious liabilities in the case of a long lease or a badly built property.

Furnished Dwellings

When furnished houses, flats or apartments are let, there is at common law an implied undertaking on the part of the landlord that the premises are reasonably fit for human habitation. This undertaking amounts to a condition of contract and if broken entitles the tenant to repudiate the contract and leave the premises or to stay in residence and sue for damages.[1]

Weekly and Monthly Tenancies

The obligations to repair in the case of periodic tenancies of less duration than from year to year were defined in *Warren v. Keen* (1954) 1 QB 15; (1953) 2 All ER 1118. It was decided that a weekly tenant is under no implied obligation to put and keep premises in repair. His only duty is to use the premises in a tenant-like manner. If the house falls into disrepair through fair wear and tear or lapse of time, or for any reason not caused by him, the tenant is not liable to repair it.

Useful examples of tenant-like manner were given in this case. For instance, "the tenant must take proper care of the place. He must, if he is going away for the winter, turn off the water and empty the boiler. He must clean the chimneys and also the windows. He must mend the electric fuses. He must unstop the sink when it is blocked by his waste. In short he must do little jobs about the place which a reasonable tenant would do. In addition, he must, of course, not damage the house, wilfully or negligently; and he must see that his family or guests do not damage it; and if they do he must repair it."

In lettings on weekly and monthly tenancies, although the landlord is generally responsible for all repairs, it is desirable to ensure that the tenant shall take care of the premises in his occupation. West[1] believes that the following agreements are suitable:

(1) The tenant shall keep the premises clean and in good condition (damage by fair wear and tear and fire excepted) and replace immediately all cracked and broken glass.
(2) Leave the premises in as good condition as the same are now in, reasonable wear and tear and accidents by fire or act of God excepted.

Tenancies from Year to Year

In the absence of express agreement, the liability for repairs of a tenant from year to year is doubtful. It has been said that he must keep the buildings wind and watertight, but there are no cases which indicate clearly the scope of his obligations. It is certain that a tenant from year to year is not liable for deterioration resulting from fair wear and tear.

West[1] has described how in the case of short leases, on yearly or three yearly tenancies, the agreement normally imposes only slight liability on the tenant for repairs. For instance, the tenant may be required to keep repaired at his own expense window and door fastenings, interior surfaces and glass to windows. The tenant shall deliver up the property at the expiration of the lease in as good repair and condition as at the commencement (reasonable wear and tear and damage by fire excepted).

Occupation Leases

Occupation leases of properties for 7, 14, 21 or more years generally contain covenants by the tenants to repair and to paint and decorate. The following covenants expressed in common legal terminology and without any punctuation illustrate typical provisions contained in this type of lease of commercial and industrial premises.

"(1) *Obligations to repair*

From time to time and at all times during the said term substantially to repair cleanse paint maintain and amend the demised premises and fixtures therein and the walls fences vaults roads sewers and drains in on or under the demised premises and the appurtenances thereof and to keep the same so repaired cleansed painted maintained and amended and free from industrial rubbish and waste materials and such part of the land (if any) as is coloured blue on the said plan for and suitable for the parking of vehicles only and the adjacent land (if any) coloured green on the said plan in a clean and tidy condition and suitably maintained as a landscaped area as laid out and as planted by the Landlord. And the demised premises so painted repaired cleansed maintained amended and kept as aforesaid at the expiration or sooner determination of the said term quietly to yield up unto the Landlord together with all additions and improvements of a permanent nature and not movable made thereto in the meantime and all fixtures of every kind in or upon the demised premises or which during the said term may be affixed or fastened to or upon the same except Tenant's trade fixtures or fittings.

(2) *Exterior painting*

In the year commencing the First day of   One thousand nine hundred and   and in every third year and in the last year of the said term however the same may be determined thoroughly to prepare and paint the outside wood and metalwork of the demised building and all additions thereto with two coats at least of best high gloss or bituminous or metallic paint (or other paint approved by the Landlord) where usually or previously so painted and thoroughly to prepare and paint all outside stonework and cement rendering surrounds or features with two coats of best stone paint (or other paint approved by the Landlord) where usually or previously so painted.

(3) *Interior painting*

In the year commencing the First day of   One thousand nine hundred and   and in every fifth year and in the last year of the said term however the same may be determined thoroughly to prepare and paint all the inside wood and metalwork of the demised building and all additions thereto with two coats of best high gloss or bituminous or metallic paint (or other paint approved by the Landlord) where usually or previously so painted and thoroughly to wash prepare stop bring forward and paint with two coats of best washable emulsion paint (or other paint approved by the Landlord) all surfaces usually or previously so treated."

A simpler form of general covenant often takes the form illustrated by West[1] and can read as follows.

"At all times during the said term to keep the premises including all fixtures and additions thereto in good and substantial repair and condition and deliver up the same in such good and substantial repair and condition to the lessor at the expiration or sooner determination of the said term."

Requirements of the Housing Acts

The *Housing Act 1957* (section 6) requires that certain houses shall be reasonably fit for habitation at the commencement of the tenancy and be maintained in that condition throughout the tenancy. The Act does not apply where the tenancy is for three years or more and the tenant has undertaken to do the repairs. Neither does it apply to houses purchased or requisitioned by the local authority for temporary housing.

The Act does not exempt the tenant from carrying out any repairs at all. Where the tenancy agreement is silent on repairs, the landlord is responsible for the repairs necessary to keep the house fit for habitation, but the tenant may have to do any additional repairs required to maintain the house in a state of tenantable repair.

The factors to be considered in deciding whether a house is fit for habitation are listed in the Act. They encompass defective ceilings, broken sashcords, defective steps inside the house, infestation with vermin and recent infectious disease.

The *Housing Act 1969* amends the factors to be considered to embrace repair, stability, freedom from damp, internal arrangement, natural lighting, ventilation, water supply, drainage and sanitary conveniences, and facilities for the preparation and cooking of food and for the disposal of waste water. At common law, the tenant alone could sue as he is the party to the contract, but the *Defective Premises Act 1972* now provides a remedy for visitors to the premises.

The landlord is given the power to enter the house at any reasonable time to view the state of repair provided that he gives 24 hours' written notice.

The *Housing Act 1961* (section 32) imposes further repairing obligations on landlords in respect of leases of dwelling houses, including flats, granted for a term not exceeding seven years. The implied covenants by the lessor comprise:

(1) to keep in repair the structure and exterior of the dwelling house, including drains, gutters and external pipes; and
(2) to keep in repair and proper working order the installations in the dwelling house:
    (i)   for the supply of water, gas and electricity, and for sanitation; and
    (ii) for space heating or heating water.

The County Court can, with the consent of the parties concerned, make an order to exclude or modify these obligations if it considers it reasonable to do so.

In *Campden Hill Towers Ltd. V. Gardner* (1977) 1 All ER 730, the Court of Appeal had to consider what was meant by the exterior of a dwelling house for the purpose of the 1961 Act. In this case, in the lease of a flat, the landlord had specifically excluded any part of the outside walls or roof, and made a demand for a service charge for the repairs to outside walls. The tenant refused to pay, contending that the service charge was not recoverable because it related to repairs which were properly within the landlord's implied obligation under section 32.

The landlord argued that the outside walls were not part of the demise and thus he was not liable for the cost of repair. The lessor's appeal from the County Court was refused as the wall was part of the particular dwelling. This liability was extended to the roof of a top floor flat by the Court of Appeal in *Douglas-Scott v. Scorgie* (1984), where it was held that the roof could be regarded as part of the structure or exterior of a flat, although a borderline case might be a top floor flat with an uninhabited attic above it.

The landlord's obligation to repair the structure and exterior refers to the structure and exterior of each individual flat, not the structure and exterior of the whole block of flats. As the Court of Appeal has stated in both cases, it is a question of fact and degree in each and every case whether some part of the building is to be regarded as the structure and exterior to any particular flat. It is clear that the structure and exterior need not be part of what is let to the tenant.[2]

The *Defective Premises Act 1972* imposes on a landlord who has covenanted to maintain or repair 'a duty to all persons who might reasonably be expected to be affected by defects in the state of the premises'. This duty is 'to take such care as is reasonable in all the circumstances to see that such persons are reasonably safe from personal injury or from damage to their property caused by a relevant defect'. The landlord's duty arises as soon as he knows or ought to know of the defect.

With leases of multi-storey flats, the question often arises as to who is responsible for the repair of the common parts in them, such as lifts and stairways, where there is no covenant relating to them in the lease. The House of Lords considered this question in *Liverpool City Council v. Irwin* (1977) AC 239, and held that there was an implied obligation on the landlord to repair the common parts.

Shopping Premises

A RICS shops working party[3] considers that the traditional repairing liability appropriate to the High Street shop can be used for shopping centres only where the building layout is simple and the responsibilities are capable of easy definition between the tenants. In these cases a repairing covenant could read 'to keep the demised premises in good and substantial repair at all times during the said term'.

However, where there are shared services and more particularly shared service areas and car parks, or the building is complex or unusual in design or layout, then it does become more difficult to determine those parts which are to be the individual tenant's responsibility and for him to maintain and insure them separately.

At the other extreme is the covered shopping centre, where the tenant is responsible only for the repair and maintenance of the interior of his shop and possibly for the shop front and fascia, with the landlord responsible for everything else. This is the most suitable arrangement where the various occupancies may be overlapping or integrated under a common roof, and it is not practical to deal with their repair in a piecemeal way.

Where there are various parts of the building shared by the tenants and overall responsibilities accepted by the landlord, the shops front on to a non-adopted pedestrian area and are serviced through private access ways and yards, then the whole approach to repair and maintenance requires careful consideration. It is advisable for the landlord to assume responsibility for all except the internal repair of the shop. The tenant normally wishes to retain control of his shop front so that he can maintain it to his own desired high standard. The desirability of operating a system of planned maintenance is evident, including future improvements, modernisation and eventually replacement.[3]

The working party[3] has described how, in a modern covered shopping centre, the lines of demarcation are difficult to establish. The landlord already recovers from tenants a large proportion of the total cost of the items of repair, maintenance and insurance of the common areas of the building and the common services (lighting, air conditioning, heating, cooling, ventilation, lifts, escalators, fire precautions, security and communications systems) and other associated services, as well as the cost of operating these services. The recovery from the tenant of the cost of repairs to the outside of the shops could conveniently also be treated as a service. All these costs and service charges will affect the level of rent which a tenant is prepared to pay.

Covenants to Repair

The nature and scope of the obligation to repair may be expressed in a variety of ways. The form of words is not particularly important as long as their meaning is clear. For example, the tenant may covenant 'to repair the premises and to yield them up in good and substantial repair and condition', or he may undertake 'to keep and leave them in good and tenantable order and repair' or perhaps 'to well and substantially repair uphold and keep them'. All these covenants express similar intentions. In all cases the tenant must repair the premises, keep them in repair and give them up in repair at the termination of the lease.

When drafting a covenant, adequate attention must be paid to the nature and condition of the premises at the commencement of the lease. Leases often operate for very long periods and changing social and economic factors are likely to affect the leased property. Hence property first leased in 1920 will contain many features which are now outdated. When the tenant carries out repairs, he is under no obligation to modernise the premises and bring them up to date. He is liable only to keep them in a satisfactory state of repair.

Meaning of Repair

To repair implies that the structure, fixture or installation is rendered fit to perform its proper function. Repair often involves the replacement of a part but it cannot be extended to cover complete rebuilding. There are several cases to support this view.

For example, in *Lister v. Lane and Nesham* (1893) 2 QB 212, the tenants under a seven year lease covenanted to 'well sufficiently and substantially repair, uphold, sustain, maintain. . .amend and keep the demised premises'. The house was at least 100 years old and before the end of the lease one of the walls was bulging outwards, and after the end of the lease was condemned by the district surveyor as a dangerous structure and was demolished. The landlord claimed from the tenants the cost of rebuilding the house which he alleged was necessary because of the tenants' neglect to repair following the serving on them of a notice to repair. The bulged wall could be repaired only by pulling down and rebuilding the entire house, since it was built off a rotten timber platform resting on 17 ft (5.18 m) of mud.

The Court of Appeal upheld the decision of the lower court. Lord Esher M.R. stated "If a tenant takes a house which is of such a kind that by its own inherent nature it will in course of time fall into a particular condition, the effects of that result are not within the tenant's covenant to repair. . . .He has to repair that thing which he took; he is not obliged to make a new and different thing,. . . . and he is not liable under his covenant for damage which accrued from such a radical defect in the original structure."

The decision in a later case – *Sotheby v. Grundy* (1947) 2 All ER 761 – was based on the earlier judgement. It was held that the tenant of a house on a 99 year lease could not be liable for the construction of new foundations to replace inherently faulty ones.

A recent case has helped to clarify the liability to repair of landlords and tenants respectively, where the damage results from inherent defects. In *Ravenseft Properties v. Davstone* (1980) QB 12, a partial collapse of the exterior stone cladding of a leasehold block of flats occurred, resulting from inherent faulty construction. The landlords undertook remedial work, replacing the defective cladding and inserting new ties and expansion joints at a cost of £55 000, on a building valued at £3m. The landlords' action succeeded on the grounds that the work related to a subsidiary part of the demised premises and in this situation the question of the tenant's liability under a general covenant to repair is one of fact and degree. The proportion of the cost of remedial works to the cost of rebuilding the whole structure is likely to be a relevant factor.[1]

However, it becomes a different situation if the tenant undertakes to pay all outgoings charged on the leased property. In *Stockdale v. Ascherberg* (1904) 1 KB 447, a tenant took a house on a three year lease and agreed to pay all outgoings in respect of the premises. During the tenancy, the landlord, on an order of the sanitary authority, reconstructed the drainage system of the house. It was held that the tenant was liable, under his agreement, to repay the landlord the cost of the work.

A tenant, who has expressly covenanted to repair, is however, liable to rebuild the entire premises if they are destroyed by his act or neglect, or by fire or other accident (*Bullock v. Dommitt* (1796) 6 TR 650 and *Redmond v. Dainton* (1920) 2 KB 256), unless he is excused by the express provisions of the lease or by statute. For example, prior to the emergency legislation of 1939, a tenant who had covenanted to repair was liable if the premises were destroyed by a bomb in an air raid.

Standards of Repair

Many phrases are incorporated in leases which attempt to indicate various standards of repair, such as 'substantial repair', 'good tenantable repair', and 'good and substantial repair'. These phrases would appear to have little real meaning and the word 'repair' would probably be adequate. 'Good tenantable repair' has been defined as 'such repair as having regard to the age, character and locality of the house, would make it fit for the occupation of a reasonably minded tenant of the class who would be likely to take it' (*Proudfoot v. Hart* (1890) 25 QBD 42).

It is necessary to qualify this definition. It has been contended that where a neighbourhood has seriously declined during the period of the lease, the tenant need repair only to the extent necessary to bring the house up to the new debased standard. Thus a house formerly in a prosperous neighbourhood which has become virtually slum property need be repaired only to a standard that would satisfy the new type of tenant, who would probably be 'satisfied with just a roof over his head'. This contention was not accepted in the case of *Calthorpe v. McOscar* (1924) 1 KB 716, where a long lease of 99 years was involved. The decision in this case was that the definition referred to the time when the lease was first granted and not to the time when it terminated.

In general the tenant under a covenant to repair must do such repairs as are appropriate for the building, having regard to its age and character at the time of granting the lease, and he must replace any parts rendered defective by lapse of time or the action of the elements.

Fair Wear and Tear Excepted

A tenant is frequently made liable for repairs 'fair wear and tear excepted'. This means that the tenant will not be liable for disrepair resulting from the normal actions of the elements, such as wind and rain, or to normal use by the tenant. He would not therefore have to repair steps worn by use, sashcords broken in use, or tiles which have slipped from the roof. He will, however, be liable for exceptional damage caused by the elements, such as hurricanes or floods, and for damage caused by the improper use of the building such as over-loading the upper floor of a warehouse, since such damage cannot arise from fair wear and tear.

In *Terrell v. Murray* (1901) 17 TLR 570, a tenant leased a house for a short term of years. The lease contained no covenant to repair during the term, but a covenant to deliver up the premises at the expiration of the term in as good a state of repair as they were at the commencement of the lease, reasonable wear and tear excepted. The tenant was held not liable for painting the exterior, repointing the brickwork or for repairing part of the kitchen floor affected by dry rot.

In *Citron v. Cohen* (1920) 36 TLR 560, a tenant covenanted to repair the interior of the premises, reasonable wear and tear excepted, and the landlord covenanted to keep the exterior in repair. The tenant was held not liable for damage to the interior caused by a leaking external rainwater pipe which the landlord had failed to repair after notice.

A further interpretation of the fair wear and tear exception was given in *Regis Property Co. Ltd v. Dudley* (1959) AC 370; (1958) 3 All ER 491, where the tenant was required to keep the house in good repair and condition, reasonable wear and tear excepted. It was held that if any want of repairs was alleged and proved, the onus was on the tenant to show that it came within the exception. Reasonable wear and tear was defined as the reasonable use of the house by the tenant and the ordinary operation of natural forces. It was further held that the tenant is bound to do such repairs as may be required to prevent the consequences flowing originally from wear and tear from producing others which wear and tear would not directly produce. An exception does not exempt the covenantor from liability for consequential damage.

Shared Responsibility

The responsibility for repairs under a lease is often shared between the landlord and tenant. Frequently the landlord is made expressly liable for external repairs and the tenant for internal repairs. Sometimes the landlord accepts liability only for structural repairs.

Whether a certain repair is external or internal is often a matter of interpretation. External covers any repairs affecting the fabric of the building, such as missing roof tiles, cracked or blocked gutters and defective windows. The term external is not, however, confined to components exposed to the elements and includes a party wall.

Where the tenant alone undertakes specific repairs, it does not signify that the landlord is responsible for the remainder. He must be made expressly liable by the lease. When the responsibility is divided, the tenant must undertake the repairs for which he is responsible whether notified of the defects or not. He is on the premises and should be aware of the defects.

The landlord, however, should be notified of defects for which he is responsible. If he has been notified and disregards the notice, he will be liable not only for repairing the defects, but also for any consequential damage resulting from his disregard of the notice. For example, if he has been notified of missing roof tiles and fails to take appropriate action, he will be liable not only to replace the tiles but also for any damage to the ceiling and decorations below.

Additional Buildings

The liability of a tenant to repair buildings which have been erected subsequent to the demise will depend largely on the wording of the covenant. The covenant may expressly cover the new buildings but, even if it does not, a covenant 'to repair' will cover all buildings both old and new alike, as annexations to the land become part of the land.

A covenant to repair the 'said demised premises and all additions thereto' would render the tenant liable to repair a garage which he has erected. Where, however, the covenant is 'to repair the demised buildings', it will apply only to buildings existing at the time the lease was granted. A covenant 'to repair and to yield up the demised premises and additions and fixtures thereto' will deprive the tenant of his customary right to remove tenant's fixtures, as discussed later in this chapter.

Liability for Damage by Fire

At common law, apart from express agreements, the tenant was not liable to rectify damage to or rebuild the demised premises after accidental fire, storm, tempest, flood or other act of God. However, the *Fires Prevention (Metropolis) Act 1774*, which was of general application, provided that no action could be maintained against any person in whose house or other building a fire should accidentally begin. This provision was, however, without prejudice to any contract between landlord and tenant. Hence where a fire is accidental and the tenant has not contracted to rebuild, he is not obliged to do so.

Similarly, the landlord need not rebuild unless he has covenanted to repair, and even if the tenant has covenanted to repair, except after a fire, it does not imply that the landlord will rebuild even if the premises were insured. The 1774 Act does, however, enable anyone interested in the building, such as the tenant, to require that an insurance company, who normally would have an option to reinstate or pay the damage, lay out the sum due towards rebuilding.

Although the house is destroyed the tenant is still liable to pay rent. It is therefore customary to insert insurance provisions in leases. The covenant normally provides for the following:

(1) who is to insure;
(2) the amount to be insured;
(3) an obligation by one party to produce receipts for premiums to the other; and
(4) an undertaking to expend any sum received from the insurance in rebuilding.

If the landlord insures it is usual to provide for payment of rent to cease unless the building is rebuilt within a specified time.

A covenant by the landlord to rebuild or reinstate the premises in the same condition as they were before the fire does not, unless so worded, make him responsible for damage to additions made by the tenant.

Painting

Paint clauses in repairing covenants are normally restricted to leases for considerable periods. Sometimes it forms part of the repairing covenant, whereas in other cases it comprises a separate covenant. Their main aim is to ensure preservation of the structure, but they also serve to compel the tenant to maintain the decorations to an acceptable standard, having regard to the class of property.[1]

Painting sometimes presents a problem of classification since it can serve two purposes. It may be undertaken to preserve woodwork and metalwork from decay which could be classed as a repair or it may be used solely for purposes of decoration to improve appearance and comfort. Some painting will serve both purposes and thus constitutes decorative repair.

Leases often contain a covenant whereby the tenant 'will in every third year paint all external woodwork and metalwork, and in every sixth year paint all internal woodwork, metalwork and other surfaces, now or usually painted and also decorate such parts of the premises as are now plastered'. In practice it is better to state the particular years in which the work is to be done and to provide for it to be undertaken in the last year of the term.

Under such a covenant the tenant's obligations are clear and no distinction need be drawn between painting for preservation and for decoration. Where the covenant is not specific and the tenant is responsible for repairs, then he must do such painting as is necessary to protect woodwork and metalwork from decay.

However, his obligation may go further and include decorations. He must then decorate according to the character of the house and the neighbourhood. In the absence of express covenants in the lease, he is not obliged to put on paint or paper of the same quality as that on the walls when he took possession and neither is he bound to paint or repaper at the end of his lease and leave the house in the same state of decoration as when he entered it.

In *Dickinson v. St Aubyn and Another* (1944) All ER 370, a landlord granted a lease for seven years with an option to the tenant to determine it at the end of five years. The lease contained a covenant to paint in the last quarter of the said term and other covenants referred to the end of the tenancy. It was held that the said term in the painting covenant meant the seven year term and, since the tenant terminated the lease in the fifth year, he had no obligation to paint. This case highlights the care that is needed in drafting covenants so that the landlord's intentions are clear.

In *Proudfoot v. Hart* (1890) 25 QBD 42, a case described earlier in the chapter, the Court of Appeal decided that the phrase 'good tenantable repair' implied that the tenant must do such repapering, painting and whitewashing as would be necessary to satisfy a reasonably minded person of the class likely to take the house. In more recent times emulsion paint has replaced whitewash and distemper. In this context, there is no real difference between 'tenantable repair' and 'good tenantable repair'.

Rights of Entry

The landlord, having granted exclusive possession of the demised premises to the tenant, cannot enter unless he is given authority to do so under the terms of a lease or a statute. If he then wishes to carry out his obligations to repair or to satisfy himself that the tenant is doing so, the lease should contain a covenant permitting him to enter. In a weekly tenancy a covenant to this effect will readily be implied, and when he has covenanted to repair a licence to enter at a reasonable time to inspect the premises will also be implied. A right of entry by the landlord to inspect the premises and carry out repairs is a common provision in a lease, but this does not excuse the tenant from notifying defects where this is required by law.

Remedies for Breach of Contract to Repair

Nature of Remedies

The grant of a lease creates a contract and the non-performance or non-fulfilment of covenants, express or implied, constitute a breach of contract for which remedies are available to the parties. These remedies comprise damages, injunction and specific performance.

Damages consist of a monetary award which will compensate the plaintiff for the damage he has suffered. This remedy would be appropriate where the landlord has accepted responsibility for external repairs which, after notice, he has failed to carry out and, in consequence, damage has been caused to the tenant's furniture, furnishings and decorations. Claims for damages may also be brought by the landlord against the tenant for breach of various covenants relating to the repair of the premises, such as a covenant to put the premises into repair, a covenant to keep the premises in repair, or a covenant to leave the premises in repair.

Injunction is an order by the Court prohibiting a party to a contract from doing some act which he has covenanted not to do. This remedy is available only for negative covenants such as where the tenant has undertaken not to carry on a trade on the premises but subsequently does so.

Specific performance is a decree by the Court which compels a party to a contract to carry out what he has specifically undertaken to do. It will not be granted where damages would be an adequate remedy nor in contracts for personal services. A landlord or tenant who refused to carry out repairs for which he was liable could be made to do so by a decree of specific performance, although, in the majority of cases, damages would be a more suitable remedy.

A good example of the operation of specific performance occurred in *Jeune v. Queen's Cross Properties* (1974) ch 97, where the landlord was in breach of his covenant to repair, when a balcony collapsed and was not reinstated. The Court ordered specific performance of the covenant in relation to the balcony, considering this a more 'convenient' course than an award of damages.

Measure of Damages

Where the tenant has covenanted to repair and has failed to do so, the amount of damages which can be recovered by the landlord is affected by whether the lease is still operative or has terminated.

(1) If the lease is still operative, the action is for breach of the covenant to repair, and the damages must compensate the landlord for the damage he has sustained. This is the amount by which the reversion has depreciated in the open market, and will be affected by the length of the unexpired period of the lease. The depreciation will be the difference between the value of the reversion with the building in repair and its value with the building in the condition of disrepair.

(2) If the lease has terminated, then the action will be for breach of covenant to repair or for breach of the covenant to deliver up the building in a proper state of repair. The damages will be the cost of putting the building into the state of repair required by the lease. For example, in *Maddox Properties Ltd. v. Davis (1950)* 155 EG 155, the Court of Appeal accepted the cost of repair as evidence of damage to the reversion and awarded the cost of those repairs to the landlord.

Forfeiture

Forfeiture of a lease will arise when a tenant has committed a breach of any of the covenants in the lease and the lease contains provision for re-entry. However, the Court is empowered to give relief to a tenant threatened with forfeiture. Furthermore, where the breach of covenant relates to matters other than non-payment of rent, the landlord must serve a notice on the tenant before he can re-enter. The landlord must then allow a reasonable time to elapse before taking further action, to allow the tenant to comply with the terms of the notice. If the tenant fails to do so, the landlord may enforce his right of re-entry, but the Court may still give relief to the tenant subject to certain conditions.

Statutory Relief

The Leasehold Property (Repairs) Act 1938 and the Landlord and Tenant Act 1954 These two Acts impose further restrictions on the enforcement of repairing covenants. At one time, landlords could serve a tenant under a long lease with a notice of dilapidations, requiring repairs to be carried out when the lease still had many years to run. This particularly benefited speculators who bought dilapidated houses cheaply and, by enforcing forfeiture for breach of covenant to repair, obtained the remainder of the term at no cost. These Acts were designed to remedy this abuse.

The Acts apply to every tenancy for a fixed period of not less than seven years of which at least three are unexpired. A landlord who serves a notice of repair may now be met by a counter-notice from the tenant claiming the protection of the Acts. The landlord is then prevented from taking action to enforce his right of re-entry, or to recover damages, except by leave of the Court.

Leave will not be given by the Court unless the landlord can prove that the repairs are necessary to:

(1) prevent substantial diminution in value of the premises;
(2) meet the requirements of a statute or byelaw;
(3) protect the occupier where he is not the tenant of the landlord; or
(4) enable the work to be done at much less cost then if it was delayed.

The tenant cannot, however, claim protection where he has accepted the liability of putting the premises in repair on taking possession, and the provisions do not apply to agricultural holdings.

The Landlord and Tenant Act 1927   This Act provides that damages for breach of a repairing covenant, whether express or implied, cannot exceed the amount by which the value of the premises has been reduced by reason of the breach. Furthermore, where the tenant has undertaken to put the premises into repair and to deliver them so repaired at the end of the lease, and he fails to do so, no damages may be recovered if the premises are to be demolished or to undergo such structural alterations as would render the repairs valueless.

The Law of Property Act 1925   Section 147 of this Act provides relief against a notice to carry out internal decorative repairs. When such a notice is served on the tenant, he may apply to the Court for relief and, if the Court considers the notice unreasonable, it may relieve him wholly or partially from liability for such repairs. The Court will have regard to the unexpired length of the lease.

This section will not, however, apply to the following:

(1) where the liability to repair arises under an express agreement to put the property in decorative repair and the agreement has never been performed;
(2) to decorations necessary to put or keep the property in a sanitary condition or to preserve it;
(3) to any statutory liability to keep the house reasonably fit for human habitation, such as required by the Housing Acts; and
(4) to any covenant to yield up the building in a specified state of repair at the end of the lease.

Party Walls

The boundary or dividing wall separating two properties may be situated entirely on one of them and the owner of the adjacent property then has no rights over it. Usually, however, each of the neighbouring owners has certain rights over the wall and it is then known as a party wall.

There is no statutory or other definition of a party wall but it may fall into one of the following four categories.

(1) A wall of which the two adjoining owners are tenants in common. In this case each owner is entitled to the use and enjoyment of the whole wall. Neither owner is entitled to remove it nor entitled to prevent the other from enjoying any part of it, such as by covering the top with broken glass or using one-half of it to support the roof of a garage.

(2) A wall divided longitudinally into two portions or strips, one portion belonging to each adjoining owner. This would mean that either owner could remove his half and leave the structure incapable of standing alone, provided that he does it in a proper and careful manner.

(3) A wall divided longitudinally into two halves, with each half being subject to cross easements of support in favour of the owner of the other half. In this case one owner could not remove his half because the other has a right of support by it, and vice versa. However, neither owner is under an obligation to keep his half in repair so that the other can enjoy the easement, although each is entitled if he so wishes to repair the other's half. This is the most common type of party wall and the *Law of Property Act 1925* converted all walls of the first type into this category.

(4) A wall belonging entirely to one of the adjoining owners but the other owner has an easement or right to have it maintained as a dividing wall. The owner of the wall can deal with it as he pleases provided that he maintains it as a party wall.

The legal presumption was that a wall came within the first category when the exact line of the boundary could not be identified. Since 1926 it is assumed to be in the third category. If the wall is built entirely on one owner's land then the presumption is that the wall is his. If it is built on the boundary so that substantially half of it is on one owner's land and half on the other's, then the wall comes in either the second or third categories. However, these are only legal presumptions and may be rebutted by proving facts inconsistent with them. There is no presumption in favour of an easement and the owner claiming it must prove that it has been acquired.

The position relating to party walls has been modified by statute applicable to some parts of the United Kingdom, particularly London. *The London Build-*

ing Acts (Amendment) Act 1939, Part VI, defines the rights and duties of adjoining owners and prescribes a code for serving notices relating to the execution of works, recovery of expenses and arbitration. The definition of a party wall given in the Act is wide enough to cover walls in sole ownership.

Two cases are now briefly described to illustrate the type of problems that can arise in connection with the repair of party walls. In *Jones v. Pritchard* (1908) 1 ch 630, the owner of a house with fireplaces and flues on the external flank wall sold a divided moiety, amounting to one-half of the wall, to the purchaser of the adjoining land, who erected a house on it. Subsequently cracks developed in the wall, permitting smoke from the flues of the newer house to enter the sitting rooms of the older house. An action was brought claiming an injunction restraining the defendant from causing a nuisance by smoke and for damages, but the action failed.

It was held that subject to the grant of the easements as to the use of the wall, the owners of the divided moieties of the wall may respectively deal with such moieties as they please. Apart from negligence or want of reasonable care and precaution, neither party is subject to any liability to the other in respect of nuisance or inconvenience caused by exercise of the rights and easements granted or reserved, including the use of the flues. The owner of a servient tenement is not bound to execute any repairs necessary to ensure the enjoyment of the easement by the owner of the dominant tenement. Furthermore, neither party is subject to any liability if, by reason of natural decay or other circumstances beyond his control, his half of the wall falls down or otherwise passes into such a condition that the easement thereover becomes impossible or difficult to exercise. Each party would, however, be entitled to repair the other's half of the wall as was reasonably necessary for the enjoyment of the easement, and this right was reaffirmed in *Bond v. Nottingham Corporation* (1940) ch 429.

In *Sack v. Jones* (1925) 1 ch 235, two houses, A and B, were separated by a party wall, house B being at the end of a row. The external flank wall of house B appeared to have subsided and was alleged to have put a strain on the party wall between A and B, which had fractured. Both owners had joint easements of support in the party wall. The owner of house A brought an action claiming that the owner of house B was bound to give lateral support to house A and to maintain the stability of the party wall, and asked for an injunction to restrain the owner of house B from permitting the party wall to be deflected or from allowing house B to be a nuisance.

It was held that there was no proof that the flank wall of house B had fractured the party wall and caused damage to house A, but, even assuming that the plaintiff had succeeded on the facts, there was no cause of action. The plaintiff had no case in respect of easements, and the servient tenement (house B) was not obliged to keep his premises in repair, although the owner of the dominant tenement (house A) could enter on the servient tenement to repair it. It was further held that the defendant had used house B only in the way in which it was intended and had not committed any nuisance.

Dangerous Structures

The law relating to dangerous structures is contained in the *Public Health Act 1936*, as amended by the *Public Health Act 1961* for districts outside London, and in the *London Building Acts (Amendment) Act 1939* for London.

Procedure in London

Where a structure is reported as dangerous, the Council (being in 1985 either the Greater London Council or the Common Council of the City of London) will request the district surveyor to survey the property. If he certifies the structure as dangerous, the Council will shore it up and take any other necessary steps to render it safe for the time being. It will also serve on the owner a dangerous structure notice which will require the owner to take down, repair or otherwise secure the structure to the satisfaction of the district surveyor. If the owner disputes the notice, he may appoint an independent surveyor who, together with the district surveyor, will appoint an arbitrator. The two surveyors jointly report to the Council and any dispute is referred to the arbitrator who must report his decision within 14 days.

If the owner does not comply with the notice, the Council may seek an order from a court of summary jurisdiction. If the owner still defaults after this order has been made, the Council itself may execute the necessary work. The Council is then entitled to demand payment from the owner of the cost of the work and, if it is not paid, the Council may recover by proceedings in a court of summary jurisdiction.

Procedure Outside London

The Public Health Act 1936, as amended by the *Public Health Act 1961*, provides that if a building is dangerous because of its condition or by reason of overloading, the local authority may apply to the Magistrates Court for an order. Under the 1961 Act, the local authority can, after notice, carry out the necessary works where a building is prejudicial to health or a nuisance, or is seriously detrimental to the amenities of the neighbourhood.

The order relating to a dangerous building may require the owner to carry out the work necessary to remove the danger or, if he so chooses, to demolish the building. If the danger is due to over-loading, the order may restrict the use of the building until the court is satisfied that any necessary works have been carried out. Where amenities are concerned, the local authority may require the repair and restoration of the building or, if the owner chooses, its demolition and the removal of the debris.

If the owner fails to comply with the order within the time specified, the local authority may carry it out in such manner as it thinks fit and recover the expenses reasonably incurred from the owner. Where the local authority is satisfied that immediate action is necessary, it may shore up or fence off the building and recover the cost of so doing from the owner.

Liability for Charges by Local and Other Authorities

Heavy expenses may be incurred in complying with the requirements of local and other authorities. In most cases notices to do the necessary work are served on the owner and he will then be liable to carry out the work, but there will be other cases where owners or occupiers will be required to carry out work to the premises.

Where the local or other authority has served a notice on the landlord or tenant requiring certain work to be carried out, the question of who will ultimately pay for the work will be largely determined by the terms of the lease. If the required work consists of repairs, then the party responsible will be the one who, by the terms of the lease, has the duty to carry out those repairs. If the party without liability pays in the first instance, he may recover from the other.

Where the work ordered to be done is not work of repair, such as the installation of a new drainage system or the making up of a road, then the statute under whose provisions the work was done will determine the liability in the first instance. A tenant may, however, have entered into a covenant 'to pay all rates, taxes, assessments and impositions whatsoever' or 'to pay all rates, taxes, duties, assessments and outgoings.'. These covenants contain wide ranging liabilities and it has been held that the word 'outgoings' includes the cost of reconstructing the house or the drainage systems under the orders of the local authority (*Stockdale v. Ascherberg* (1904) 1 KB 447). On the other hand, liability for the payment of 'rates, taxes and assessments' would include only charges of a recurring nature and not exceptional demands which enhance the permanent value of the property.

Fences

West[1] has described how the word 'fence' includes any physical barrier between two tenements such as a wall, palings, posts and rails, posts and wire, a hedge, bank or ditch. Frequently a fence consists of a combination of two or more of these barriers, such as a ditch, a bank and a hedge.

Ownership

Where the boundary of land is determined by reference to a fence, the precise line of the boundary is often difficult to establish. This difficulty cannot always be resolved by reference to the title deeds, although these form the primary source of evidence. Apart from the title deeds there is a presumption that the fence is on the land of the person who made it to avoid trespassing on his neigh-bour's land. In the case of a bank and ditch, it is presumed that these belong to the person whose land is on the bank side of the ditch. The owner would norm-ally dig the ditch up to the edge of his land, throw the earth on to his land to form the bank and then often plant a hedge on the bank. This presumption applies only to artificial or man-made ditches, as where a stream separates land in different ownerships, the boundary will follow the centre line of the stream. Likewise a conveyance made by reference to an Ordnance Survey map may take the boundary to the centre of the hedge.[1]

With wooden fences, ownership is normally decided, in the absence of evi-dence to the contrary, by the position of posts and stays which are presumed to be on the land of the owner. If the fence is paled, the fair face of the boards form the boundary and are exposed to the neighbour. Aldridge[4] has pointed out that a wall or boarded fence may be set back slightly from the boundary to avoid trespassing on erection or maintenance.

Repairs

The liability to repair fences follows the same principles as for other repairs. The tenant is liable in the absence of an express covenant assigning the duty to the landlord. The obligation to repair fences arises from the tenant's duty to preserve the boundaries of the demised premises.

The obligation is one between landlord and tenant. It is not an obligation which is owed to the adjoining neighbour unless there is an agreement between them that one or other will fence. This procedure applies even where two occu-piers have a common landlord and the tenants have entered into covenants to repair fences. If A and B are tenants of X and A has covenanted to keep in repair the fence separating the fields occupied by A and B, A's obligation to fence is a duty owed to X and not to B. It follows that if B's cattle stray on to A's land because of A's neglect to repair the fence, A can maintain an action of trespass against B. B cannot excuse himself by pleading A's neglect to repair the fence, for this is a duty which A only owes to X and not to B.

A landowner may acquire an easement that his neighbour shall fence the boundary between their properties, or shall maintain a hedge along it. Aldridge[4] examined two contrasting cases which both attempted to link easements and repairs of fences. In *Hilton v. Amkesson* (1872) 27 LT 519, evidence that the owners of a piece of land had repaired a certain fence for 50 years was held insufficient to establish an easement, in the absence of any evidence that they did so as a result of an obligation. In contrast in *Lawrence v. Jenkins* (1873) LR 8 QB 274, evidence that a fence had been repaired by one party and his predecessors for 40 years was held sufficient, when combined with the fact that for the last 19 years they had done so on notice from those claiming the benefit of the easement.

Fixtures

Nature and Significance

The general rule applying to fixtures is that whatever is attached to the soil becomes part of it. When a building is erected and components are permanently fixed to it, the soil, building and the components all constitute land. Roofing tiles, rainwater goods and windows stacked on the site prior to incorporation in the building are chattels, but once they are fixed they become land, as land in the legal sense includes buildings. However, if the building is demolished and components become severed from it, then they revert to chattels. The question of whether an object is a fixture and as such part of the land or alternatively a chattel is important, for it may determine its ownership. Thus a wash basin purchased and fixed by a tenant in the demised premises may cease to be his property and become that of his landlord. Furthermore, a purchaser of land or premises may claim that he has purchased all the objects fixed to the land, which includes buildings, at the date of his contract to purchase.

Purpose of Annexation

The stringent rules of common law have been modified considerably and the Courts now direct most attention to the purpose of annexation. If the object is an integral part of the building, such as a chimney, water service pipe or bath, then it will be a fixture as it is obviously intended to be a permanent feature of the building.

If, however, the object is installed temporarily and is not intended to become an integral part of the building or land, then although it may be necessary to fix it so that it can be used and enjoyed, it will remain a chattel. Thus if a tenant purchases a gas fire and attaches it to a gas service pipe, but does not intend the appliance to be a permanent feature, then it remains a chattel and the tenant can remove the fire when he leaves the premises.

Distinction between Fixtures and Chattels

The legal position is clearly defined in Halsbury's *Laws of England* — "Whether a chattel has been so fixed to land or buildings as to become a fixture depends on the purpose of annexation, and if the chattel can be removed without irreparable damage to the premises, neither the method nor the degree of fixing, nor the amount of damage that would be done to the chattel or the premises by the removal, affect the question save in so far as they throw a light upon the object and purpose of the annexation.

If the purpose was for the permanent and substantial improvement of the land or buildings, the article will be deemed to be a fixture, but if it is attached to the premises merely for a temporary purpose, or for the more complete enjoyment and use of it as a chattel, then it will not lose its chattel character and it does not become part of the realty."

Degree of Annexation

The degree of annexation is certainly not the ultimate test of a fixture. Indeed there may be no annexation at all and yet the object may become a fixture, as was the intention of the person so placing it. Thus stone blocks arranged to form a dry wall are not fixed but they constitute a fixture, whereas similar blocks stacked on a patio would be chattels. Much therefore depends on the intention of the person who fixed the object, although the degree of annexation may be important in providing evidence of intention. If the object is so fixed that its removal would cause irreparable damage to the land or building, then it could be inferred that it was not intended to be removed and has accordingly ceased to be a chattel.

It has been held that an electric lighting generating machine fixed by bolts to the floor, a conservatory on brick foundations, and seats bolted to the floor of a cinema were fixtures as they had become an integral, substantial and even necessary part of the building. Statues, figures, vases and garden seats have also been considered fixtures as they were essentially part of the design of the house and grounds although standing independently by their own weight.

In contrast, valuable tapestries fixed to a framework nailed to the walls and surrounded by a moulding, a collection of stuffed birds in cases fixed to walls, panelling, window blinds and ornamental fireplaces have been held not to be fixtures, as fixing them to the walls was no more than was necessary for their enjoyment as chattels. These examples illustrate that the demarcation line can be rather blurred on occasions.

Right to Remove Fixtures

Once it is established that an object is not a fixture it can be removed by the person who brought it on to the land or his successors in title. Whereas if it is a fixture it becomes the property of the landowner and cannot be removed.

There are, however, some important exceptions to this rule and many objects which are undoubtedly fixtures, such as shopfittings, are regarded as tenants' fixtures and can be removed by them. Despite the use of this terminology, the legal title to the fixtures is in the landlord until the tenant exercises his right to remove them. The tenant may remove these fixtures during his tenancy or within a reasonable time after it has ceased. For instance, where the tenancy is a weekly one, the tenant cannot reasonably remove tenant's fixtures within a week, and he will be allowed a reasonable time to remove them after the termination of the tenancy. Once however that time has elapsed he loses his right to remove them, and the landlord acquires an absolute title to the fixtures.

Tenant's Fixtures

The main categories are as follows:

(1) *Trade fixtures.* These are fixtures attached to the premises for the purpose of the tenant's trade or business. Showcases, public house fittings, petrol pumps fixed to underground storage tanks, vats, boilers and engines attached to the ground have all been adjudged tenants' fixtures and as such can be removed by them. Shrubs planted by a market gardener have been similarly classified, although similar shrubs planted for ornamental purposes are deemed to be ordinary fixtures and cannot be removed by the tenant.

(2) *Ornamental and domestic fixtures.* These encompass articles which can be removed without damage to the premises or where the damage is easily repairable. There are many borderline cases. For instance a heating appliance in a conservatory could be construed as part of the structure and is therefore a fixture. Alternatively it could be regarded as an accessory and as such would be removable.

No general rule can be prescribed but it is evident that an essential part of a structure cannot be removed so that what remains is worthless. Thus the shelves of a conservatory may be easily removed and the structure can still function as a conservatory, although possibly not so effectively. If, however, the glass was removed the conservatory could not perform its normal purpose. Hence the glass is an essential component of the structure and, as such, constitutes a fixture.

(3) *Agricultural fixtures.* By virtue of the *Agricultural Holdings Act 1948*, a tenant may remove fixtures that he has annexed either during or within two months after the end of the tenancy, provided the following conditions are fulfilled:
  (i)   one month's notice must be given to the landlord;
  (ii) the tenant must have fulfilled all his obligations under the tenancy;
  (iii) no avoidable damage must be done in the removal and any damage done must be made good; and
  (iv) the landlord may retain the fixtures if he serves a written counter-notice on the tenant and pays a fair price for them. The price is assessed on their value to the incoming tenant.

Liability for Damage to Persons and Property

Liability of an Occupier towards Visitors

The Occupiers' Liability Act 1957 abolished the distinction between invitees and licensees. An invitee is a person who visits the premises on some matter in which both he and the occupier have an interest, whereas a licensee comes for some purpose of his own but with the occupier's permission. The occupier now has the same duty towards all his *lawful* visitors.

The Act made no reference to the occupier's liability to trespassers, who enter the premises without authority, and hence common law rules still apply. His liability is well illustrated in *British Railways Board v. Herrington* (1972) AC 877, where a small boy crossed a broken-down fence, which the Railways Board had failed to repair, and was severely injured by an electric railway line. In upholding his claim for damages, the House of Lords decided that the standard by which an occupier would be judged in relation to trespassers was that of 'ordinary common humanity'. Although this was not so burdensome on the occupier as the common duty of care under the *Occupiers' Liability Act*, it nevertheless required him to take such steps as could reasonably be expected of a 'conscientious, humane man with his knowledge, skill and resources'.

An occupier must exercise a common duty of care to any invitees and licensees who may visit his premises. The 'common duty of care' is to take such care as the circumstances reasonably demand to ensure that the visitor is reasonably safe for the purpose for which he is invited or permitted by the occupier to be on his premises. In assessing what degree of care the circumstances demand, the occupier must look to the experience, capabilities and propensities of the visitor. He must, for example, be prepared for children to be less careful than adults.

On the other hand, he may reasonably expect that anyone who visits the premises in the exercise of his calling will appreciate and guard against any special risks likely to be associated with it. A builder may, for example, assume that the building control officer, architect, quantity surveyor and sub-contractors will all be familiar with the dangers normally encountered on a building site.

The Unfair Contract Terms Act 1977 has modified the position of the occupier whereby he cannot contract out of his business liability under the *Occupiers' Liability Act 1957* for the death or personal injury of lawful visitors on his land, resulting from negligence.

If a visitor disregards a warning given by the occupier and he suffers injury, the occupier will be absolved from liability only if he can show that his warning, if observed, was sufficient to make the visitor reasonably safe. A board with the word 'danger' on it would generally be inadequate, since the visitor would not know the nature of the danger or from where it might come. It may be necessary for the occupier to rope or fence off the dangerous area or to post a person to keep visitors away.

The Occupiers' Liability Act 1957 contains a provision of considerable significance for all contractors. Where the occupier has employed a competent contractor to carry out works of construction, repair or maintenance on his premises, he may be absolved from liability towards a visitor who has been injured because of the faulty execution of the work. The occupier must have taken all reasonable steps to satisfy himself that the contractor was competent and the work properly done. This provision reaffirms the legal principle that a contractor's duty is not confined to his obligations under the contract. He has a general duty to take reasonable care to prevent injury to *all* persons who he may reasonably expect to be affected by his work.

Landlord's Obligation to Repair

Where a landlord is under an obligation to repair or maintain premises occupied by his tenant, he will have the same duty towards the tenant's visitors in respect of dangers resulting from his neglect to repair as if he were the occupier. For example, if a landlord has received notice of a cracked and dangerous ceiling which he ignores, he will be liable to a person visiting the tenant who is injured by falling plaster.

This liability is clearly defined in the *Occupiers' Liability Act 1957*, which provides that in respect of dangers arising out of any default in his repairing obligations, the landlord shall be liable to visitors as if he were the occupier. It is, however, normally a prerequisite that notice of the need for repair must have been given to the landlord by the tenant.

The landlord's obligation was further widened by the *Defective Premises Act 1972*, which provides that a landlord who is under an obligation, however imposed, to repair owes a duty of care to all persons likely to be affected to see that they are not injured by defects in the property. The duty applies to all defects which constitute a breach of the repairing obligation or which would amount to a breach if the landlord had notice. Furthermore, the duty arises not only when the landlord knows of the disrepair, but also when he ought reasonably to have known.

Where a landlord lets out the premises in parts, as for example in flats or offices, he frequently retains control of the entrance hall, stairways, lifts, corridors and roof, and is regarded as the occupier of those parts of the premises. Accordingly he owes a duty of care to all persons lawfully using the parts that he retains. The duty owed to tenants may be covered by express terms in their contracts, but if not, a liability will be implied by law.

Premises Adjoining the Highway

It is a public nuisance to permit premises abutting on the highway to fall into a ruinous state, and any person who lawfully uses the highway and is injured because of the state of disrepair is entitled to damages. In *Tarry v. Ashton* (1876) 1 QBD 314, the plaintiff was injured because of the fall of a heavy lamp which projected over the highway. The occupier was held liable, although he was unaware of the dangerous condition of the lamp and had actually had it repaired a short time before.

Another relevant case will serve to illustrate the principle involved. In *Slater v. Worthington's Cash Stores (1930) Ltd* (1941) 1 KB 488, the plaintiff was on the pavement looking in the window of the defendants' shop, when a large fall of snow from the roof of the premises injured him. The snow resulted from unusually severe storms and it could have been removed at any time within four days before the accident. The defendants took no steps to remove it, nor did they give any warning to persons using the pavement, nor prevent any persons from using it. The premises were occupied by the defendants under a lease whereby the landlord was liable for external repairs, and the roof and guttering were in satisfactory condition.

The plaintiff claimed damages on the grounds of negligence and nuisance, and the defendants claimed that they were not liable by reason of the landlord's covenant to repair the roof. The Court held that:

(1) the accumulation of snow on the roof amounted to a public nuisance and that, as the defendants had done nothing to abate it, they were liable for the damage suffered by the plaintiff;
(2) there was a duty on the defendants to safeguard members of the public using the pavement from the danger caused by the snow, and as the defendants had not fulfilled that duty, they were liable for negligence also;
(3) even if the storms were to be regarded as an act of God, which was doubtful, that defence could not help the defendants, as it was not the storms that directly caused the plaintiff's injury, but the fact that the snow had not been removed; and
(4) the landlord's covenant to repair did not absolve the defendants from liability.

These decisions were subsequently unheld in the Court of Appeal

A person who makes an excavation on his land close to the highway and fails to fence it, commits a public nuisance and will be liable to anyone who falls into it while using the highway. The excavation will not be a nuisance if it is so far from the highway that the injured person must have been a trespasser.

Liabilities of Owners and Occupiers

Where a person has been injured by reason of the defective state of the premises it is the occupier who, *prima facie*, is liable to him. However, the owner will be liable if he has covenanted with the tenant to repair and the injury has been caused by his failure to do so. He may also be liable if he has let the premises in a ruinous condition and the tenant has not agreed to repair them.

Neither the owner nor the occupier will be liable for a nuisance caused by a third party without his knowledge and which he has no reasonable opportunity to rectify. Thus in *Baker v. Herbert* (1911) 2 KB 633, some boys broke part of the railings of an empty house, causing the area to become a danger to anyone using the street. The plaintiff, a child, fell through the gap in the railings and was injured. It was held that the owner was not liable, since he was unaware of the broken railings and insufficient time had elapsed to enable him to have known and undertaken the repairs.

Liability to Neighbours

It is a trespass for an occupier to permit anything to encroach on his neighbour's land. Thus branches of trees which overhang or roots which undermine adjoining land constitute a trespass, although it is a frequent occurrence.

An occupier is liable to an adjoining occupier for injuries arising out of the dilapidated condition of his premises, in a similar manner to his liability to those who use the highway. Thus where defendants had allowed a rainwater pipe to become choked with leaves so that the water which it should have carried overflowed, it was held that they were liable for damage caused to the plaintiff's premises by the water discharged on to them.

An occupier of land continues a nuisance if with knowledge or presumed knowledge of its existence he fails to take any reasonable means to stop it, although he has ample time to do so. In the absence of such knowledge no liability exists. This is illustrated in *Wilkins v. Leighton* (1932) 2 ch 106, where many years previously the defendant's predecessor in title built a house, which resulted in pressure on a retaining wall which subsequently collapsed. It was found that the defendant neither knew or could have known that the house constituted a nuisance to the neighbour's property, and therefore that he was not liable.

References

1. W. A. West. *The Law of Dilapidations*. 8th edition. P. F. Smith. Estates Gazette (1979)
2. B. Denyer-Green. Landlord's responsibility for roof repairs. *Chartered Surveyor Weekly* (23 February 1984)
3. RICS Commercial Property Committee. Model repairing clauses. *Chartered Surveyor* (December 1981)
4. T. M. Aldridge. *Boundaries, Walls and Fences*. Oyez Longman (1982)

8 Schedules of Dilapidations

This chapter is concerned with the preparation of schedules of dilapidations both during a lease and at its termination. The relevant precautions to be taken and procedures to be followed are examined in some detail, followed by typical schedules to illustrate their format and content.

Taking Instructions

Instructions may emanate from the client or his solicitor or other agent, or the appointment may be for a person to act as an independent surveyor or arbitrator, taking representations from the two parties or their surveyors. As described in chapter 1, in connection with structural surveys, all instructions should be clear and concise and confirmed in writing, and include the basis of assessment of fees and expenses, which can be costly, to avoid any possible disputes in the future. On occasions it may be possible only to quote the fees and expenses on a provisional basis in view of the uncertain nature and extent of the work. It will help the client considerably if he is notified periodically of the amount of fees and expenses incurred to date.

Schedules of dilapidations can take one of two forms – either an interim schedule listing defects that require remedying to comply with the repairing covenants during the term of a lease, or a terminal schedule that details the necessary items of repair at the end of the lease. A RICS Guidance Note[1] identifies the need for the landlord's surveyor to advise his client where it seems likely that the measure of damages could be restricted significantly by statute, case law, schedule of condition or other relevant factors.

The surveyor should obtain and carefully examine the tenancy documents and check their authenticity and reliability as far as practicable. For instance, it may contain dates that are inconsistent and the plans accompanying the lease may be poor quality copies, inaccurately coloured and deficient in notation. The surveyor should take immediate steps to resolve any ambiguities in the documents or instructions from the client, and it is always desirable to determine the landlord's intentions for the future use of the building as this could affect the surveyor's approach.

The RICS Guidance Note[1] points out that the surveyor's duty to a lay client may be more exacting than that to an instructing solicitor. For example, he may need to warn a tenant to serve a counter-notice within a statutory period. A surveyor often gives advice on relatively simple legal matters but the client should be referred to a solicitor for advice on more complex issues.

Leases

The RICS Guidance Note[1] explains why it is essential for the surveyor who is instructed to prepare a schedule of dilapidations to obtain the complete lease and associated documents, or reliable copies suitably coloured where appropriate. A recital of the repairing covenants alone could be insufficient for any of the following reasons.

(1) Identification of the demised premises is often shown on a plan or described in a schedule or in an appendix to the lease.

(2) It is necessary to determine whether the landlord and/or his representative have right of access to the premises, the period of notice required and the method of serving it, where appropriate.

(3) Schedules of condition and schedules of fixtures and fittings can form an integral part of the title document. As described in chapter 6, where a lease contains a provision that the tenant is not required to keep or leave the property in better condition than at the commencement of the lease, a schedule of condition should be agreed between the parties and be attached to and form part of the lease, to record the condition of the property.

(4) Leases frequently impose an obligation to comply with statutory requirements, which can extend or vary the repairing covenants, as described in chapter 7. The surveyor must also be on the lookout for any outstanding statutory notices.

(5) A tenant may be in breach of a covenant either to alter the premises or to refrain from doing so.

(6) Covenants are frequently incorporated in leases to prohibit the tenant making structural alterations.

(7) Licences may be granted permitting the carrying-out of alterations subject to subsequent reinstatement.

(8) Where there are several tenancies in a building, each purporting to have a full repairing obligation towards part of the demised premises, it will be necessary to examine all the leases to ensure that the covenants are the same.[2]

Repairing Covenants

There are three categories of repairing covenant which the surveyor will need to consider in dealing with dilapidations:

(1) express covenants specifically expressed in leases and agreements;
(2) implied covenants such as to use the premises in a tenant-like manner and not to commit waste; and
(3) covenants imposed by statute, such as by section 32 of the *Housing Act 1961*.

As described in chapter 7, express covenants should be precise and unambiguous in their meaning and scope. They normally apply to the demised areas only, except where there are express references relating to common or other parts in multiple occupied properties.

The surveyor must consider the effect of current case law on the interpretation, application and implementation of dilapidations clauses, as it can be very significant. Some of the more important cases are outlined in chapter 7 and many more are detailed by West.[2] He must also have regard to statutory liabilities as, for example, Regulations made under the *Health and Safety at Work Act 1974* and the *Factories Act 1961*, as their effect can be substantial, particularly in the case of industrial buildings.

The surveyor should check on any obligations of either the landlord or tenant relating to the notification of the need to repair. This may be required by the lease or agreement or be a common law requirement. He will also need to consider the dilapidations implications of the failure of the other party to carry out repairs for which he is liable.

Where the dilapidations clause contains ambiguities, the surveyor should endeavour to interpret the clause by reference to the probable original intentions of the parties to the lease, the client's solicitor and relevant case law. The surveyor must also take account of the impact and effect of such common phrases as 'at the termination of the tenancy, the premises are to be delivered up in as good a state as they were at the commencement' and 'fair wear and tear excepted'.

As illustrated in chapter 7, repairing covenants can be drafted and interpreted in a variety of ways, and extensive guidance is provided by appropriate case law. A covenant 'to keep in repair' implies 'to put in repair' if premises are not in repair at the commencement of the tenancy. Another important consideration, as described in chapter 7, is the fact that premises need not be delivered up in the same condition as they were at the beginning of the lease, particularly in the case of long leases. The age, character and locality of the premises and their immediate future use must all be taken into account.

When preparing schedules of dilapidations, the surveyor needs to place a comparable modern interpretation on clauses defining work finishes and standards which are no longer applicable. It should also be borne in mind that in the case of a lease of a dwelling house for a term of less than seven years, section 32 of the *Housing Act 1961* imposes obligations on the landlord in respect of repairs to the structure, exterior and services, regardless of the express requirements of repairing covenants, as described in some detail in chapter 7.

Alterations and Improvements

Alterations are often self-evident while improvements may be the subject of conjecture, and so alterations are normally deemed to include improvements to avoid possible controversy. Alterations can often be identified by the existence of one or more of the following conditions:

(1) differences in construction materials;
(2) materials which are not commensurate with the age of the building;
(3) parts of the premises which are clearly associated with the tenant's trade or occupation, such as the presence of hoisting tackle in a light engineering workshop; and
(4) plans and records, where available.

Where the lease prohibits alterations or is silent on the matter, the landlord's surveyor should provide for reinstatement. However, the tenant's surveyor may wish to oppose this requirement where no injury has been caused to the reversion. Changing the nature or character of the demised premises by means of alterations is technically waste, as defined in chapter 7. If, however, it could be classified as ameliorating waste then the tenant's surveyor is entitled to contest the requirement to reinstate, provided that the alteration is not in conflict with any statutory provisions.

Licences and consents relating to alterations should be examined to determine the form of approval and to establish its authenticity. Alterations which have been carried out badly or vary significantly from the approval should be included in the schedule of dilapidations as works of amendment or repair.

Alterations which fall within a conditional reinstatement clause should be included in the terminal dilapidations schedule. On the other hand, alterations which have been carried out as a result of a statute which permits or requires such alterations, such as section 165 of the *Housing Act 1957* or statutory notices under the *Fire Precautions Act 1971*, should not form part of the claim unless they are poorly executed or in disrepair.

Interim Schedules of Dilapidations

Preparatory Work

The first step is for the surveyor to take instructions from his client and confirm them in writing. The instructions must set out clearly the purpose for which the survey and subsequent activities are required. The surveyor should then obtain a copy of the lease and all associated documents and examine them thoroughly. He should also check to see whether any repair notices have been issued and, if so, what action has been taken.

If the lease has several years to run, a decision will be needed as to whether to recommend to the client that a notice to repair should be served or an interim schedule. To assist in making this decision, the surveyor normally makes a preliminary inspection of the property to ascertain the general standard of repair. Where a considerable number of properties are involved and the surveyor is unfamiliar with the area, he would be wise to widen his inspection to encompass the neighbouring area, to become familiar with the age and character of the buildings and the standard of repair that might reasonably be accepted in the locality. In particular he will note the existence of any derelict sites and properties that are boarded up and will make enquiries to determine the reasons for this.

Following the preliminary inspection, a visit to the local planning office is advisable, to ascertain the local authority's plans for the future development of the area. The surveyor would also wish to know from the local authority whether there are any notices, listings or designations affecting the property or the area in which it is situated. The surveyor will now be in a position to advise the client as to the best course of action. If the local authority is contemplating demolishing the property then no action may be required, while if the property is being reasonably well managed and maintained then a notice to repair would be more appropriate than an interim schedule of dilapidations, in the way of a timely reminder that routine maintenance matters require attention.

Inspection

When acting for the landlord, the surveyor will need to make the necessary arrangements for inspection strictly in accordance with the provisions of the lease, and when acting for the tenant the necessary appointment is often made at the time of accepting instructions. When carrying out the inspection on behalf of the landlord, the tenant and/or his representative may wish to accompany the surveyor, although it is preferable that he carries out the inspection alone and uninterrupted. When acting for the tenant, the surveyor will take with him the schedule prepared by the landlord's surveyor, examine all the items of disrepair listed in the schedule and insert his own notes beside them. The inspection can be carried out alone or accompanied by the landlord.

The inspection of the property normally follows the same procedure as adopted for structural surveys and described in chapter 4. It is essential that a methodical and logical approach is adopted to avoid any omissions and to enable others to follow the order of items on site without having to retrace their steps unnecessarily and to easily assimilate the schedule. The same procedure or format should be used for each property, floor or room.

Any dangerous or potentially dangerous parts of the property which are liable to cause damage or injury should be recorded and the landlord, tenant and occupants advised of any immediate action that should be taken.

It is often advisable to provide a sketch plan, with a north point inserted on it, showing the layout of the building(s). All the rooms and other component parts should be numbered or lettered for purposes of identification, and additional information such as building names, room names and occupancy use can be added for reference purposes.

Some surveying practices use prototype schedules with numbered standard clauses. The notes on a site inspection are normally made on a room by room basis and the notebook entry inserted against each part of the premises consists of a series of numbers denoting the standard clauses. Only in cases where the standard clauses do not apply would the surveyor need to describe the defect or remedial work in full.

It is not customary on an initial inspection to engage specialists to inspect and advise on the condition of mechanical and electrical engineering services, unless the need for this has been clearly established, and even then the client's approval should be obtained.[1]

Essential Formalities

When acting for the landlord, the surveyor should check the schedules of condition and of fixtures and fittings if they exist. He should also check all licences and/or permissions that have been issued in connection with the demised premises, and also any structural or other alterations carried out, with or without approval, and take them into account when subsequently preparing the interim schedule of dilapidations. In the case of alterations undertaken without the landlord's approval, instructions should be obtained from the client as to whether retrospective approval from the local authority should be sought.

When acting for the tenant, the surveyor should check the schedules of condition, and of fixtures and fittings, where they exist, against the schedule of dilapidations and note the discrepancies on the schedule of dilapidations.

The Schedule

The RICS Guidance Note[1] describes how the schedule should clearly and concisely list the defects and the necessary repairs. Generalisations should always be avoided in the drafting of schedules although they lack the comprehensive nature of the wording contained in specifications and bills of quantities. It is also advisable to avoid the inclusion of items which are unlikely to result in a significant reduction in the value of the reversion, as they might well be counterproductive in generating avoidable opposition from the tenant.

West[2] has shown how an interim schedule is often drafted to set out in general terms the work required to comply with the tenant's obligations to repair. Typical entries being 'repoint defective areas of brickwork' and 'overhaul all missing, slipped and damaged roof tiles', without specifying the precise locations. The surveyor must ensure that damages for the breach of the repairing covenant will not exceed the amount by which the value of the reversion in the premises will be reduced. For example, broken, missing or decaying structural components must diminish the value of the reversion, but internal decorations and minor repairs generally do not. Furthermore, the practical effect of the *Leasehold Property (Repairs) Act 1938* is to exclude all internal decorations and minor repairs where three or more years of the lease remain unexpired.

The inclusion of measured items in an interim schedule and their subsequent pricing is not usual as the primary purpose of the interim schedule is to ensure that work is carried out to comply with the repairing covenants of the lease.

Terminal Schedules of Dilapidations

A terminal schedule of dilapidations can be prepared during the last three years of a lease but is most commonly produced during the last year.

Preparatory Work

It is necessary for the surveyor or other professional to confirm instructions, obtain all relevant documents and make appropriate enquiries as previously described for interim schedules. The client's needs and future intentions must be clearly identified. In the case of extensive properties or complex leases, it is usually advisable to consult a solicitor. Where appropriate, the local authority should be contacted to determine whether there are any notices, listings or designations affecting the property or the neighbouring area.

Inspection

A general reconnaissance of the locality and the premises is required in the manner outlined for interim schedules. It is important that adequate notes and dimensions are recorded to permit detailed schedules to be prepared and costed, and hence the survey needs to be more thorough than for an interim schedule. The underlying causes of defects must be identified. For example, buckled bitumen felt roofing may result from distorted or decaying supporting timbers, while fractured concrete paving may be caused by the action of tree roots or leaking drains or water services. Where rising damp is encountered or the repair of specialist equipment is required, it may be deemed advisable to obtain a specialist's report and estimate.

It cannot be over-emphasised that the inspection must be performed in a logical and methodical manner, and to avoid recording detailed information on matters which cannot form part of the claim, or where a claim for damages will be limited to the reduction in value of the reversion which will be considerably less than the cost of repair. A sense of proportion must be maintained throughout.

The RICS Guidance Note[1] identifies the need to make detailed sketches or take photographs of items which will be the subject of future consideration away from the site, and this will certainly assist with evidence in Court. Survey notes and dimensions must be carefully and neatly recorded as they may subsequently be referred to in a Court or by an Official Referee.

Where the property is occupied at the time of the inspection, any tenant's fixtures should be noted, together with any remedial work that may be necessary following their removal. With services, a decision will be required as to whether specialist advice is required, particularly in the case of mechanical and electrical services in large complex buildings.

The Schedule

The schedule will almost inevitably form the basis of extensive negotiations and must therefore be produced accurately, concisely and positively. It must be correctly titled and refer specifically to the property, the landlord and the tenant by name, and include the dates of the lease document and the inspection of the property.[1]

It is advisable for the schedule of dilapidations to be prepared in such a way that a Scott schedule can readily be prepared subsequently from it, should this become necessary. It should follow a logical sequence and contain adequate locational references, specifying clearly the rooms or other parts of the premises in which the defects occur, so that the reader can easily find his way around the schedule. Ideally paragraphs should be numbered for ease of identification and each paragraph should describe clearly and succinctly the nature of the necessary repairs. The schedule should, as far as practicable, be written in non-technical terms so that its contents can be readily understood by a layman, with ample headings and sub-headings to guide him through the document.

Prior to the preparation of the schedule of dilapidations, any specialist reports, such as those covering services or structural matters, should be collated with the surveyor's site notes and all recommended repairs included in the schedule.[1]

When determining the nature and scope of the repairs considered necessary, the surveyor must have full regard to the requirements of the lease and the repairing covenants in particular. He must also take into account the class, age and location of the property and so determine the reasonable requirements of the class of tenant likely to occupy the premises. This could have a considerable influence on the requirements relating to decorations.

West[2] has described how there are often instances where the original components in an old building are no longer obtainable, such as door furniture, certain types of floor finish, roof tiles and eaves gutter sections. In these situations, the schedule should read 'repair or renew', as the term 'reinstate' would be inappropriate. A final check list often includes such items as door keys, sashcords, radiator keys, electrical fittings and thermal insulation. Full attention must also be paid to external works such as drives, paths, boundary walls, fences and gates.

The renewal of items can normally be enforced only where repairs are impracticable. Alterations and additions need special investigation, as described earlier in the chapter, to ascertain whether the approval of the landlord was sought and obtained, together with any necessary Building Regulations and Town and Country Planning approvals. It is necessary also to consider their effect on the value of the reversion and the possible need for their demolition and the making good of the premises.

Normally in a schedule of dilapidations, the internal items are more fully detailed and dimensioned, as it is necessary to detail the defects in each room in turn, followed by the staircase, hall and corridors or passages. Externally, it is normally sufficient to include general clauses covering the roof, rainwater goods, walling, windows and doors, and paintwork for each elevation. It will be appreciated that a surveyor may subsequently be required to produce his site notes and dimensions in Court or before an Official Referee, who is a High Court Arbitrator specialising in building disputes.

In preparing the schedule of dilapidations, the surveyor will be concentrating on specifying the defects which require making good to comply with the repairing covenants. He will not, however, be describing in detail the method of carrying them out and, in this respect, a schedule of dilapidations differs significantly from a specification. It is not unusual for one or two copies of the schedule to be typed on bill of quantities paper to assist in the pricing of the items of repair.

When preparing schedules of dilapidations for properties in multiple occupation, the surveyor will commence by identifying from the lease the demised area of a particular tenant and the common areas used by all tenants. He will also identify any licensed works and other structural alterations undertaken in the demised premises, and any equipment and plant, wherever located, which will be removed by the outgoing tenant, and make allowance for any subsequent making good. The surveyor must ensure that no service charge items appear in the schedule of dilapidations when he is acting on behalf of the tenant.

Service of Schedules and Notices of Repairs

Notices and counter-notices under the *Law of Property Act 1925* and the *Leasehold Property (Repairs) Act 1938* may be served by a surveyor but the more usual and normally the best practice is for them to be served by solicitors. The surveyor will frequently be called on to give advice as to the liabilities, obligations and rights of the client, be he landlord or tenant, and he should exercise the utmost care and skill in the performance of this duty. The principal ramifications are described in chapter 7.

Interim schedules of dilapidations can be served on the tenant by means of a simple notice requiring the repairs to be carried out within a reasonable period or the time can be specified in the lease. The notice can be served by the landlord, managing agent, surveyor or a solicitor. A reasonable period for execution of the repairs would normally be in the order of three months but this could be excessively long if urgent repair work such as the remedying of dry rot is required, and too short if extensive and intricate structural work has become necessary.

Service of the terminal schedule of dilapidations frequently takes place before the end of the lease, as practical difficulties may arise if it is left too late. It is an advantage if the tenant is given the opportunity to undertake the repairs before the lease ends. Service of the schedule at or after the expiry of the lease should be accompanied by a claim for damages.[1]

Claim for Damages

Assessing the Cost of Building Work

The RICS Guidance Note[1] recommends that the itemised schedule should be costed in a manner appropriate to the size and complexity of the property. A sensible balance needs to be maintained as, for instance, it would be inappropriate to spend a long time preparing the equivalent of a bill of quantities while, at the other extreme, decorations should not be spot priced on a room by room basis unless current comparables are available. Measurements and areas are normally required to enable negotiations to take place between surveyors, and will almost certainly be necessary to justify or refute a claim in Court.

West[2] advocates that every item of work should be measured and valued to enable an inclusive figure to be stated in the claim. He describes how each item in the schedule should be priced, including 'spot prices' where unmeasured items occur. In practice the items are often abridged ones containing a combination of activities which would constitute separate measured items in a bill of quantities. It is necessary to keep a sense of proportion and the entries in the schedules which appear later in the chapter show a realistic approach.

The most costly items in claims for dilapidations frequently relate to commercial properties let on full repairing leases. For example, an industrial building may be covered with extensive areas of corrugated asbestos cement sheeted roofs and numerous intervening valley gutters, which could be in a very dilapidated state and will not be visible from ground level. In the case of old industrial buildings, the steel frameworks and roof trusses may be badly corroded and, while descaling may be sufficient in some cases, other parts may require costly replacement. Office buildings may contain large areas of metal windows which are severely corroded, and their urgent replacement may be necessary. These examples illustrate the types of problems that can face the landlord's surveyor, arising from the high cost involved in repairs and replacements, frequently accentuated by difficulty of access. The tenant's surveyor is almost certain to challenge the scope and cost of every item.[2]

The RICS Guidance Note[1] describes how composite prices can often be applied to the less complex properties as, for example, by measuring decorations over all surfaces, without making adjustments for windows, doors and similar components. The use of standard price books, such as those published by Spon, Laxton, Griffiths and Wessex, provide useful guidelines but the prices contained in them should always be used with the utmost care. There is unfortunately no such thing as an average price and costs will vary with location, amount and nature of work, and other variable factors. A knowledge of current local prices will always be a great asset in arriving at realistic prices for the scheduled items. On occasions builders' quotations have been obtained to provide a basis for pricing some of the more complex items of work, and this approach is further complicated by the fact that a schedule of dilapidations is not prepared in a form that readily lends itself to pricing by builders. The estimated cost of the work of repairs and replacements should be based on prices operative at the date of termination of the lease or any agreed extension.

Other Assessment Aspects

The surveyor acting for the landlord will record the necessary particulars and dimensions on the site, and will subsequently calculate the relevant areas and lengths from the site dimensions. He will then proceed to price each item in the schedule of dilapidations, although the tenant receives an unpriced copy.

Where scaffolding or other plant is needed for more than one item of work, the surveyor can either allocate a proportion of the plant costs to each item of permanent work or alternatively insert the plant costs in separate preliminaries items. The cost of the general builder's attendance on sub-contractors' work is normally incorporated in composite cost items.

Where the repairs are of a complex nature, they may entail supervisory costs and it will be a matter for negotiation between the surveyors as to whether these costs are to be included in the claim. Where the lease makes provision for the payment of surveyor's fees then these can form part of the claim, and the same principle applies to the fees of the landlord's solicitor. Where appropriate, Value Added Tax will be added to the surveyor's and solicitor's fees.

Where the tenant has failed to comply with the repairing covenants in the lease, the landlord's surveyor should determine whether any other loss is likely to be incurred. Typical examples are loss of rent, loss of service charges in buildings in multiple occupation and payment of rates during the period required to carry out the repairs.[1]

Procedural Aspects

Following discussions with the other surveyor, it may be necessary to recommend to the client that the only realistic approach is one of compromise because of the difficulty of clearly defining where the liability lies. It is advisable to obtain the client's approval prior to entering negotiations, to clarify whether authority is given to reach a settlement or, alternatively, to reach a 'without prejudice' agreement to recommend a figure in settlement of the claim.

The surveyor acting for the landlord may find it necessary to reinspect the property after vacation by the tenant, to determine whether any further breaches of repairing covenants have occurred.

When acting for a tenant, the surveyor should consider whether a statutory limit applies to the measure of damages that can be claimed. He will also consider carefully the reasonableness of the scheduled works and the costs ascribed to them. From the information obtained and dimensions taken he will assess the costs of each item of work included in the schedule of dilapidations.

Settlement of Claims

When a lease is nearing its expiration date, it is customary for the landlord's surveyor to prepare a schedule of dilapidations so that the tenant will have the opportunity to carry out his repairing obligations under the lease. The tenant may however prefer to settle the claim by means of a cash settlement, and will then appoint his own surveyor to advise on the extent and cost of the work listed in the schedule for which he is considered liable. On occasions the tenant may accept the landlord's assessment and pay his surveyor's fees.

Following the serving on the tenant of the surveyor's schedule and a claim for damages by the landlord's solicitor, the surveyors for the two parties will normally meet together on the site. At this meeting it should be possible to agree on the physical condition of the property and then to examine together the items making up the schedule. Where the items are reasonably straightforward, the surveyors may be able to agree the prices of the individual items and then to settle the claim for damages at one and the same meeting. Alternatively, the tenant's surveyor may wish to check all the schedule items on the site before he meets the landlord's surveyor. Unless the surveyor has authority to settle the claim without reference back to his client, he should agree the terms 'without prejudice' and then take his client's instructions.[2]

In addition, surveyors will need to check instructions, terms of the lease, relevant correspondence, any schedules of condition and/or fixtures and fittings and examine any tenant's improvements. In the case of offices and factories, tenants may have erected partitioning and false ceilings and fitted cupboards, shelves, double glazing and other components. It will be necessary to decide whether these additions and alterations contravene the lease or constitute genuine improvements. In the latter case the value of the improvements should be taken into account in the final settlement. Other factors requiring attention could include surveyor's fees, loss of rent and whether the value of the landlord's schedule exceeds the diminution in value of the reversion.[2]

However carefully the schedule is prepared, there are bound to be differences of opinion between the two surveyors as to the extent of the repair work and its probable cost. Normally the surveyors will by negotiation be able to reach a solution which is mutually acceptable, and they would be wise to inform their respective clients that they will need some latitude in their negotiations. It must always be borne in mind that inflexible and uncompromising stances can result in clients incurring further legal and other costs. Most clients prefer to settle quickly rather than have the case referred to the courts.

Where agreement cannot be reached, the clients should be advised of the areas of disagreement so that a decision can be made to accede, proceed to litigation or refer the matters in dispute to an independent surveyor, where this is provided for in the lease. A Scott schedule as illustrated in chapter 6 and later in this chapter, will generally be required by the solicitor and/or the Court. The Scott schedule is normally prepared by the plaintiff's solicitor, based on information supplied by the surveyor.[1]

It is advisable for the landlord's surveyor to warn his client or the solicitor that the amount of the claim may be challenged and, where possible, to give an indication of the possible eventual settlement figure. If this action is taken at an early stage, it will assist in obtaining the client's approval to a subsequent negotiated figure or alternatively help the client to decide whether to take legal action to recover damages. The tenant will generally welcome similar advice from his surveyor. Surveyors would be unwise to adopt an inflexible approach to the preparation and settlement of claims as it is likely to rebound to their disadvantage. Even when a surveyor has a strong case, one or two minor concessions will frequently lead to a quick settlement which is almost inevitably in the interests of both parties.

Compensation for Business Premises

With the exception of business premises, compensation is not payable to a tenant in respect of alterations. The landlord will usually give the tenant the option of removing the alteration work and reinstating the premises or leaving the altered premises in a good state of repair.

With business premises the tenant has a right to claim compensation for improvements under Part I of the *Landlord and Tenant Act 1927* as amended by the *Landlord and Tenant Act 1954*. For this right to operate, the tenant must have served notice on the landlord of his intention to make improvements, fully detailed on a plan and specification. If the landlord does not object within 3 months or the Court gives approval, then the improvement is authorised and may attract compensation. This will not however apply where the work is required to comply with statutory requirements or a contract condition, or to trade or other fixtures which the tenant has a legal right to remove.

A notice of claim must be served in accordance with the Supreme and County Court Rules. For example, where a lease ends by passage of time, the claim must be served not more than 6 months nor less than 3 months before the expiration of the lease.

The amount of compensation must not exceed the lesser of:

(1) the net addition to the value of the property which directly results from the improvement; or

(2) the reasonable cost of carrying out the improvements at the end of the lease (less the cost of any remedial work to the improvement).

In determining the increase in value of the property, regard must be paid to any intended change in use, demolition or alterations that would limit or nullify the compensation.[1]

Specimen Schedule of Dilapidations

The following example illustrates a specimen schedule of dilapidations which lists the items needing repair but does not indicate the manner in which the repairs are to be done in the way that a building specification does.

The schedule is prepared in a logical sequence starting on the upper floor of the property and working downwards. It is helpful to insert an approximate estimate of the cost against each item in the site notes, although they will not normally appear in the schedule. It may be necessary, however, to give an indication of the financial outlay involved. When producing a schedule with a view to quantifying a claim for damages, a wide right hand margin should be provided for pricing purposes. Finally, when preparing schedules the terms of the lease or agreement must be borne in mind to ensure that only dilapidations for which the tenant is liable are included.

Schedule of Dilapidations and Wants of Repair found to have accrued at 32 New Street, Ambleton
under the terms of a lease dated 1 May 1978 between P. T. Arrowsmith and H. B. Waterson
prepared by Tape and Measure, Chartered Surveyors, 6 The Rope Walk, Edmonstone.
16 May 1985

INTERNALLY

First Floor
Bedroom 1   Clean, stain and varnish margins to floor.
                 Strip, stop, prepare and hang walls with paper of similar quality to existing.
                 Cut out cracks in plaster ceiling, make good, wash, stop and twice emulsion paint.

Wash, stop and paint two coats of oil paint to all woodwork previously painted.

Renew cracked and missing tiles to fireplace hearth.

Bedroom 2   Repair defective floorboarding.

Clean, stain and varnish margins to floor.

Strip, stop, prepare and hang walls with paper of similar quality to existing.

Replace lock key and oil and adjust lock.

Reinstate broken metal fanlight stay.

Refix loose tiles to fireplace hearth.

Bedroom 3   Clean, stain and varnish margins to floor.

Strip, stop, prepare and hang walls with paper of similar quality to existing.

Cut out cracks in plaster ceiling, make good, wash, stop and twice emulsion paint.

Wash, stop and paint two coats of oil paint to all woodwork previously painted.

Renew broken sash cords to double hung sash window.

Bathroom   Repair defective flooring.

Remove projecting nails from walls and make good.

Wash, stop and paint two coats of oil paint to plastered walls.

Cut out cracks in plaster ceiling, make good, wash, stop and twice emulsion paint.

Wash, stop and paint two coats of oil paint to all woodwork previously painted.

Renew broken coat hook on door.

Reinstate cracked wash basin and make proper connections to all pipes.

Thoroughly clean stains from inside of bath.

Rewasher hot water tap to bath.

Ground Floor
Hall,   Strip, stop, prepare and hang walls with paper of similar quality
Staircase and   to existing.
Landing   Cut out cracks in plaster ceiling and soffits, make good, wash, stop and twice emulsion paint.

Clean and polish handrail and newels.

Renew three broken balusters.

Renew two defective treads.

Reinstate broken panes of glass in lead lights to landing.

Lounge   Clean and wax polish oak parquet floor.

Strip, stop, prepare and hang walls with paper of similar quality to existing.

Wash, stop and paint two coats of oil paint to all woodwork previously painted.

Reinstate broken firegrate.

Dining Room Strip, stop, prepare and hang walls with paper of similar quality to existing.

Cut out cracks in plaster ceiling, make good, wash, stop and twice emulsion paint.

Ease door to hall.

Renew defective door lock.

Refix loose casement fastener.

Reinstate two broken panes of glass in French casements and renew defective putties.

Kitchen Cut out cracks in ceiling, make good, wash, stop and twice emulsion paint.

Wash, stop and paint two coats of oil paint to all woodwork previously painted.

Reinstate missing fastener to fitted cupboard.

Wash, stop and paint two coats of oil paint to plastered walls.

Reinstate defective draining board.

Reinstate cracked white glazed tiles at back of sink.

EXTERNALLY

Building Repair defective roughcast near bathroom window.

Fabric Repoint open and defective mortar joints to chimney stack.

Reinstate defective and missing roof tiles and refix loose tiles.

Clear secret gutters to valleys.

Clear eaves gutters and downpipes and make good defective and leaking joints.

Repoint open joints to door and window reveals.

Reinstate two defective window sills.

Fences Renew defective oak gate post and fastenings.

Repair and straighten wire fence near gate.

Repair defective hanging stile to gate.

Renew length of fence near double gate.

Replace defective small gate.

Paths Replace defective areas of concrete paths.

GENERALLY

Reinstate all defective fastenings to doors and windows where not previously mentioned, oil defective locks and replace any missing keys. Examine and leave in working order all electrical fittings, ball valves and taps and clean out cisterns. Clean out all gully traps and manholes, well flush out drains and leave in satisfactory condition. Sweep all flues. Clean all glass on both sides. Clean all floors. Remove all debris and rubbish on completion of works.

Extracts from an Interim Schedule of Dilapidations

The following example contains possible extracts from an interim schedule of dilapidations which might be served on a tenant during the term of a lease, as to leave them unremedied could lead to accelerated deterioration of the property and diminution in value of the reversion.

Schedule of Repairs required to make good the Dilapidations to 62 Thorneywood Grove, Blakenham in accordance with the covenants of the lease dated 12 March 1973, between the Trustees of Joseph Grainger deceased and Goldstone Developments Ltd.
prepared by Level, Staff and Chain, Chartered Surveyors, St. Mary's Close, Manningford
20 June 1985

EXTERNALLY

Garage Roof Take down the whole of the defective roof covering and decayed timbers to the roof of the garage and rebuild as before with wall plates, rafters, collars and ridge board. Fix roofing felt, battens and matching tiles, including ridge tiles, and point verges. Fix new fascia and soffit boarding. Replace cracked cast iron eaves gutters and downpipes to match existing.

Knot, prime, stop and paint fascia and soffit boarding with three coats of oil paint and prime and paint three coats of oil paint on gutters and downpipes.

Leave the roof sound and watertight on completion.

Front Boundary Wall Take down the 12 m defective length of front boundary brick wall, excavate and grub up the foundations. Construct a new foundation and rebuild the one brick wall in matching bricks in cement mortar (1:3) with a lead-based bitumen damp-proof course. Point the wall on both faces and finish with a brick on edge coping and tile creasing to match existing.

Rear Fence Replace defective and decayed pales and decayed gravel board in close boarded fence.

Arch to Back Door Cut out defective arch and rebuild in stock facing bricks to match existing.

Front Chimney Stack Take down the defective chimney stack to roof level and rebuild as before, including the provision of a lead-based bitumen damp-proof course and make good the lead flashings, soakers, apron and gutter.

Rear Wall to House Rake out and repoint with a flush joint, to match existing, the top six courses of brickwork and all other joints which are open or defective.

Renew one cracked stone window sill.

Front Wall   Take out two decayed casement windows to the lounge and
to House   replace with similar type windows, adequately treated with suitable preservative and knotted, primed, stopped and painted with three coats of oil paint. Point around frames and make good to reveals, soffits and sills as necessary.

INTERNALLY

Typical   Take down the cracked and bulging lath and plaster ceiling, and
Bedroom (to   make out with new metal lathing and plaster to match existing
illustrate   and make good to adjoining work.
approach)   Wash, stop, seal and twice emulsion paint ceiling.

Cut out defective plaster to walls and make out in new plaster to match existing, including fair flush joint to existing.

Strip, stop, prepare and hang walls with paper of similar quality to existing.

Renew defective casement stay bars and fasteners.

Renew decayed floorboarding and joists.

Replace defective lock to door and provide keys.

Replace cracked tiles to fireplace hearth.

Wash, stop and paint two coats of oil paint to all woodwork previously painted.

GENERALLY

Clean out all gullies, traps and manholes, and rod and well-flush drains and leave in satisfactory condition.

Take off, repair and oil defective locks and other door furniture, and replace all missing keys. Renew all defective or missing window fastenings.

Examine and leave in working order all electrical and gas fittings, ball valves and taps, and clean out cisterns.

Sweep all flues. Remove all debris and rubbish on completion of works.

The surveyor, having prepared the interim schedule of dilapidations, will then normally proceed to produce a schedule of claim. The items contained in the schedule will generally be listed as abridged or composite items against which prices can be inserted.

For example, it is unlikely that the garage roof will be subdivided into individual items for roof tiling and battens, felt, double course at eaves, ridge tiles, verges, wall plates, rafters, collars and ridge board, as when preparing a bill of quantities. A more common approach is to build up a price for the roof measured on the flat plan area over all external walls and not the area on slope. The price would include the tiling, battens, felt and all roof timbers. The only additional items to be taken would be the eaves, inclusive of fascia, soffit board, eaves gutter and painting; verges, including barge boards where appropriate; and ridge.

In like manner the front boundary wall could be taken as a single linear item embracing the excavation, grubbing up existing foundations, backfill, concrete foundations, brickwork, damp-proof course, fair-face work and coping. The price for the composite item will be built up from the prices of the component parts, in a similar way to the method used in approximate estimating.

Proof of Evidence Incorporating a Schedule of Dilapidations

Background Information

Mr. P. M. Rossington was a tenant occupying premises at 15 Norfolk Street, Haversham, Hampshire for 21 years under a lease dated 21 May 1964. The tenancy agreement included a covenant to keep and leave the house and grounds in good and tenantable repair and condition, fair wear and tear excepted. The landlord, Development Enterprises Ltd, submitted to Mr Rossington a schedule of dilapidations six weeks prior to the expiration of the lease. The tenant disputed the schedule, refused to comply with it and instructed a solicitor to contest it. The landlord has commenced legal proceedings and a surveyor, Mr Jonothan Sharpe, has been instructed to prepare a proof of evidence incorporating the schedule of dilapidations and a schedule of claim for breach of the repairing covenant.[3]

Proof of Evidence

Development Enterprises Ltd v. Peter Michael Rossington.
Claim for damages for breach of repairing covenant in respect of lease of 15 Norfolk Street, Haversham, Hampshire.
JONOTHAN SHARPE
will say
Qualifications
  (1) I am a Fellow of the Royal Institution of Chartered Surveyors in practice on my own account at 28 High Street, Haversham. I have had twenty eight years' experience in the profession of a surveyor and have undertaken a large amount of work involving dilapidations throughout this period.
Subject of Appeal
EXH 1 (2) I produce the lease of 15 Norfolk Street, Haversham granted by the plaintiff to the defendant for a term of twenty-one years from 21 May 1964. (*Note*: EXH refers to exhibit.)
  (3) EXH 1 repairing covenant
       The lessee agrees at all times during the said term to keep the premises, including all fixtures and additions, in good and tenantable repair and

condition and to deliver up the same in such good and tenantable repair and condition to the lessor at the expiration or earlier determination of the said term.

EXH 2 (4) I produce a schedule of dilapidations EXH 2 and a schedule of claim
EXH 3   EXH 3 signed by me as the plaintiff's agent, showing the sum of £6340 as the cost of the items of repair therein set out and constituting the plaintiff's claim for damages.

(5) The defendant was presented with the schedule of dilapidations EXH 2 six weeks before the expiration of the lease. None of the repairs listed in the schedule has been carried out.

(6) The plaintiff claims damages for breach of covenant to keep the premises in repair in accordance with the terms of the lease EXH 1. The basis of the claim is exhibit EXH 3 (schedule of claim).

Inspections

(7) I inspected the premises on 18 April 1985, 14 May 1985 and 7 June 1985.

(8) I have examined the lease of the property directing special attention to the repairing covenant and I consider that the schedule of dilapidations EXH 2 has been prepared in strict accordance with the terms of the lease.

Evidence of Dilapidations

(9) The repair of the property has been neglected for a considerable time and I could find no evidence of any recent repairs to the premises. I submit that the want of repairs listed in the schedule of dilapidations EXH 2 is evidence that the defendant has not carried out his obligations under the repairing covenant EXH 1.

EXH 2 (10) *Schedule of Dilapidations*

A. No decorations have been carried out for a number of years and certainly not within the last year of the lease as required by the repairing covenant. The wallpaper in the hall and lounge is badly discoloured and disfigured. The paintwork to plastered walls and ceilings elsewhere is also badly discoloured. Paintwork to woodwork both inside and out has deteriorated badly, with bare wood showing in a number of places.

B. Rainwater gutters and downpipes are rusting badly with some cracked sections and leaking joints, causing damp brickwork and the growth of algae.

C. Putties to windows on the south and west sides of the house have perished and there are many cracked panes of glass.

D. Twenty-five roofing slates are missing from the south and west roof slopes, permitting some rainwater penetration into the roof space.

E. Pointing to extensive areas of brickwork on the south wall of the house has perished.

F. The WC pan in the bathroom is cracked and the flushing mechanism is defective.

G. Two manhole covers are badly cracked and corroded.

H. The close boarded fence on the southern boundary is badly decayed and in a state of collapse.

EXH 3 (11) *Schedule of Claim*   £

A.   Prepare plastered walls and ceilings and apply two coats of
    emulsion paint
        walls 550 m² @ £3.60   1980.00
        ceilings 160 m² @ £4.50   720.00
    Strip off existing wallpaper and replace with new
        80 m² @ £4.30   344.00
    Rub down woodwork, touch up with primer and apply three
    coats of oil paint
        90 m² @ £7.00   630.00

B.   Take down defective rainwater gutters and downpipes and
    replace with new
        gutters 30 m @ £10.80   324.00
        downpipes 15 m @ £15.60   234.00
    Prepare, prime and paint gutters and pipes with three coats
    of oil paint
        60 m @ £3.50   210.00
    Remove algae growth from brickwork
        5 m² @ £2.20   11.00

C.   Rake out defective putties to windows and replace with new
        40 m @ £3.80   152.00
    Hack out cracked glass to windows and reglaze
        8 m² @ £18.50   148.00

D.   Replace missing roofing slates
        25 @ £5.20   130.00

E.   Rake out joints of brickwork and repoint
        70 m² @ £8.60   602.00

F.   Replace WC suite   135.00

G.   Replace two manhole covers @ £74   148.00

H.   Replace decayed close boarded fence
        22 m @ £26.00   572.00

                                      Total   £6340.00

(12) The estimate of £6340 is based on current prices.
Date: 21 June 1985   Signed: Jonothan Sharpe

Use of Scott Schedules in Settling Dilapidations Claims

The RICS Guidance Note[1] shows very effectively how a Scott schedule can be used to settle claims for damages arising from dilapidations. Table 8.1 illustrates the process by reference to some specific examples.

    Each item of dilapidations is listed and where entries in the schedules of condition or of fixtures and fittings, prepared at the commencement of the lease, are relevant to the point at issue, they are also extracted and entered on the Scott schedule. The schedule also summarises the amount claimed in respect of

Table 8.1 Use of Scott schedule in settling dilapidations claims

Item nr	Dilapidations: particulars of work	Schedule of Condition	Schedule of Fixtures & Fittings	Amount claimed (£)	Amount (if any) admitted by defendant (£)	Plaintiff's observations	Defendant's observations	Judge's notes
	Treatment and Toilet Area							
	Gents Toilet							
4.1	Redecorate ceiling	Plastered and painted with oil paint. Paint flaking and in poor condition	—	50.00	nil	Decorations needed to deliver up premises in good condition at end of lease	Ceiling decorations in poor condition at commencement of lease. Dilapidations claim fails to take account of Schedule of Condition	
4.2	Clean ceramic tiles to walls	Glazed ceramic tiles in good condition	—	40.00	25.00	Wall tiling is very dirty	Agreed but dispute cost	
4.3	Redecorate woodwork	All woodwork gloss oil painted and in good condition	—	100.00	75.00	Paintwork disfigured and discoloured	Agreed but dispute cost	
4.4	Replace defective WC pan	Vitreous china sanitary ware in good condition	2 nr vitreous china WC pans with low level cisterns and plastic seats	90.00	70.00	One WC pan badly cracked	Make some allowance for fair wear and tear	

Table 8.1 (continued)

Item nr	Dilapidations: particulars of work	Schedule of Condition	Schedule of Fixtures & Fittings	Amount claimed (£)	Amount (if any) admitted by defendant (£)	Plaintiff's observations	Defendant's observations	Judge's notes
	Ladies Toilet							
5.1	Redecorate ceiling	Plastered and painted with oil paint. Paint in moderate condition	—	65.00	25.00	Decorations in poor condition with flaking paint	Make allowance for decorations only being in moderate condition at commencement of lease. Dilapidations claim fails to take account of Schedule of Condition	
5.2	Clean ceramic tiles to walls	Glazed ceramic tiles in good condition	—	50.00	30.00	Wall tiling is very dirty	Agreed but dispute cost	
5.3	Redecorate woodwork	All woodwork gloss oil painted in fair condition, with some dirty and scratched areas	—	120.00	35.00	Paintwork badly disfigured and discoloured	Make allowance for paintwork only being in fair condition, with deficiencies noted in Schedule of Condition at commencement of lease	
5.4	Replace one defective WC suite	Vitreous china sanitary ware in good condition, except for one crazed pan	4 nr vitreous china WC pans with low level cisterns and plastic seats	135.00	70.00	One WC pan is cracked and flushing mechanism is defective	Make allowance for crazed pan as listed in Schedule of Condition	

Table 8.1 (*continued*)

Item nr	Dilapidations: particulars of work	Schedule of Condition	Schedule of Fixtures & Fittings	Amount claimed (£)	Amount (if any) admitted by defendant (£)	Plaintiff's observations	Defendant's observations	Judge's notes
5.5	Renew damaged door and replace defective lock and door furniture	Entrance door to toilet split on shutting edge	—	50.00	8.00	Shutting edge badly split and door lock and furniture defective	Door can be repaired and main defect existed at commencement of lease and is listed in the Schedule of Condition. Liability limited to replacing door lock and furniture	
	Entrance Hall							
6.1	Redecorate ceiling	Plastered and painted with oil paint. Paint in fair condition but some small areas flaking	—	75.00	nil	Decorations needed to deliver up premises in good condition at end of lease	Decorations in poor condition at commencement of lease and dilapidations claim takes no account of Schedule of Condition	
6.2	Renew damaged section of decorative panelling	Iroko veneered plywood decorative panelling 1.22 m high, fixed to battens, in good condition	—	45.00	35.00	3 m length of panelling badly cracked and twisted	Agreed but dispute cost	

Table 8.1 (continued)

Item nr	Dilapidations: particulars of work	Schedule of Condition	Schedule of Fixtures & Fittings	Amount claimed (£)	Amount (if any) admitted by defendant (£)	Plaintiff's observations	Defendant's observations	Judge's notes
	Entrance Hall (ctd)							
6.3	Repair pair of entrance doors	Entrance doors, oak panelled and glazed in good condition	—	55.00	30.00	Both doors opened at joints and jamming. Require dismantling, cramping, rewedging and rehanging	Some allowance should be made for fair wear and tear, as defects stem mainly from normal usage	
6.4	Reinstate loose wood blocks to floor and seal and wax polish the whole floor	Iroko wood block flooring wax polished and in good condition	—	60.00	35.00	Two areas of loose blocks require re-bedding and floor surface sealing and polishing	Agreed but dispute cost	
	Treatment Room							
7.1	Redecorate ceiling	Plastered and painted with white emulsion paint. Paint rather dirty and slightly discoloured	—	180.00	nil	Decorations in poor condition and repainting necessary to deliver up premises in good condition at end of lease	Decorations in poor condition at commencement of lease and dilapidations claim takes no account of Schedule of Condition	
7.2	Repair crazed areas of terrazzo dado	Grey *in situ* terrazzo dado 1.27 m high, cracked for full height in two places	—	150.00	50.00	5 m² of terrazzo dado require cutting out and replacing to remove crazed areas	Part of crazing existed at commencement of lease and dilapidations claim takes no account of Schedule of Condition	

Table 8.1 (continued)

Item nr	Dilapidations: particulars of work	Schedule of Condition	Schedule of Fixtures & Fittings	Amount claimed (£)	Amount (if any) admitted by defendant (£)	Plaintiff's observations	Defendant's observations	Judge's notes
	Treatment Room (ctd)							
7.3	Redecorate wall surfaces above dado	Plastered and painted with oil paint, matt finish, in fair condition	—	240.00	90.00	Decorations in very poor condition, badly disfigured and discoloured	Allowance to be made for decorations only being in fair condition at commencement of lease as evidenced in Schedule of Condition	
7.4	Replace cracked terrazzo panels to floor	Black and white terrazzo laid in alternate panels about 500 × 500 mm between ebonite dividing strips, all in good condition	—	100.00	60.00	Seven panels of terrazzo badly cracked and require replacement	Agreed but dispute cost	
7.5	Replaster area of perished wall plaster	Plaster in sound condition	—	40.00	nil	4 m² of plaster is hollow and has lost its key to the wall	Plaster is basically sound and loss of key has been caused by movement in the building which is commensurate with its age	

each item by the landlord and his supporting observations, together with the amount, if any, admitted by the tenant with his observations. The judge or other appropriate person can enter his notes beside each item as it is considered.

The entries in table 8.1 show how widely the views of the two parties can diverge, and the value and significance of schedules of condition and of fixtures and fittings in coming to a decision on the amount to be set against each item. It should also be borne in mind that the tenant is normally not liable for the effect of fair wear and tear. It will immediately become apparent that the evaluation of many dilapidation items is quite difficult and that a reasonable attitude must be displayed by both parties in order to secure agreement to appropriate figures.

The two principal considerations running through these negotiations are as follows:

(1) The condition of the property at the expiration of the lease should normally be similar to that prevailing at the commencement, after allowance for fair wear and tear.
(2) The condition of the property should be acceptable to a reasonably minded tenant who would normally occupy this type of property.

References

1. Royal Institution of Chartered Surveyors. *Building Surveyors Guidance Note: Dilapidations* (1983)
2. W. A. West. *The Law of Dilapidations.* 8th edition. P. F. Smith. Estates Gazette (1979)
3. I. H. Seeley. *Building Maintenance.* Macmillan (1976)

Appendix A: Survey check list — site factors

Element	Item to check: drawing eventually at 1:1250, 1:500, 1:200
External works	
Boundaries	Defined in relation to the building line
	General orientation, aspect
	Position of benchmark (if any)
Ground surface	Level or sloping
Walls, fences and gates	Material/type
	Condition
	Heights
	Direction of swing of gates
Steps, ramps and paved areas	Material/type, condition
Natural features	Tree type, shrubs, creepers
	Condition, heights, spread, etc.
Ground condition	Liability to flooding — height above sea level (from maps), streams, evidence (marsh roads), water table
	Liability to subsidence (mining or other) — evidence (uneven ground, cracking)
	Liability to erosion (coastal)
Adjoining properties and outbuildings	Material/type
	Condition, construction/general form
	Proximity of undesirable features
Drainage	
Manholes/inspection chambers	Type, size, direction of flow of pipes
	Invert level, cover level
	Adequacy of fall, condition
Drains, gullies, vent shafts	Testing for water tightness by hydraulic, smoke, or pneumatic methods
	Material type and size

Element	*Item to check: drawing eventually at 1:1250 1:500, 1:200*
Soil and vent pipes, waste pipes	Condition Adequacy (may be necessary to excavate)
Cesspools, septic tanks, soakaways, wells, (if present)	Understand drainage pattern and ultimate disposal system
Services For all services	Main supply and position Material, adequacy, condition Protection
In addition: Water, rising main	Storage Stopcock and drain cock
Hot and cold water (if present)	Control Insulating material for hot water Degree of protection from frost
Electrical	Control and meter position Lighting and socket outlets — conduit runs, age, etc.
Gas	Control and meter position
Heating	Boiler, radiators — position, storage, capacity Fuel — gas, oil, solid fuel Pipe condition Supplementary provision/electric immersion
Telephone	Location of supply and outlet

Appendix B: Survey check list — building elements

Element	Item to check: drawing eventually at 1:100, 1:50
Roof/exterior	
Pitched, flat or mixed	Construction — hipped or gable, shape and slope, complexities, for example, dormers Main finish, hips, ridge, verge
Eaves	Construction — parapet, exposed rafters, bargeboard projection
Gutters and rainwater pipes	Materials, size Condition, adequacy Fixing, discharge points Rib, valley and parapet gutters Is the rainwater system complete? Check elevations
Flashings	Chimneys, party walls, abutments, pipes traversing, gutters
Roof interior	Construction — main beams or trusses, rafters, spacing and size
Ceiling	Ceiling joist size and spacing (if different) Material and thickness Boarded areas Condition of timbers, infestations
Flues in roof space	Rain penetration
Party or gable walls	Back of stacks Torching (if applicable)
Insulation	Material, capacity
Storage cistern	Insulation, support, age, condition, cover, presence of valves (if any)
Pipework	Material, conditions, insulation, functions (supply, downfeed, overflows)

Element	Item to check: drawing eventually at 1:100, 1:50
Walls	
External finish/material	Materials
	Condition
	Thicknesses (insulation value)
	Pointing, condition
	Decoration
	Structural system in general (mass frame or mixed)
	If stone or brick — bonding, coursing method, rubble infill, reinforcement?
Elevations	Rainwater pipes
	Particular features — balconies, bay windows, dressings, overflows
Air bricks/grilles, ventilators	Type, size, adequacy, obstructions, connected to ducts? Disused?
Foundations	Plinth
	Subsoil, type, settlement
	Defects; cracking, bulging, patches, staining
Damp-proof course	Material (slate, engineering brick, bituminous)
	Relative position to ground level
	Obstructions, malfunctioning, continuity
	Rising damp
	Pointing, condition
Windows	Type and size
	Material, finish
	Glazing, beads or putty conditions
	Sub-frame
	Lintel, arch
	Sill threshold
	Position in rebate, flashings
	How fixed?
	Direction of swing of opening
Doors	As for windows, plus — additional fanlight, single, double

Element	Item to check: drawing eventually at 1:100, 1:50
Floors	
Timber floor	Construction, joist size and direction
	Stability, deflection
	Finish, surface condition (square edged or tongued and grooved boarding)
	Infestation, damage, defects (likely points under windows, corners, near walls)
	Access traps
Stairs, timber, concrete or steel	Construction, sizes, tread and riser (pitch)
	Soffit, open or sealed
	Treads and risers, finish, nosing condition
	String, well apron, balustrade/railing
	Stability
	Deflection (movement relative to walls)
Concrete floor	Construction, screed thickness
	Position of downstand beams
	Ceiling finish and thickness (if appropriate)
	Damage
	Finish
Internal walls and ceilings	
Recesses, projections	Are projections closed chimney breasts?
	Do they occur on other floors?
Plaster	Ceiling, walls, condition, thickness, ornamentation/roses/centres
	Is it alive? (Loosely pinned to back of structure or only stuck to wallpaper?)
Finishes and trim	Woodwork condition (architraves, etc.)
	Decorative repair and standard
	Are architraves, cornices complete?
	Special decorations

Index